AF696208

Ecological Research Monographs

Series Editor
Yoh Iwasa, Department of Biology
Kyushu University
Fukuoka, Japan

The book series Ecological Research Monographs publishes refereed volumes on all aspects of ecology, including. Animal ecology Population ecology Theoretical ecology Plant ecology Community ecology Statistical ecology Marine ecology Ecosystems Biodiversity Microbial ecology Landscape ecology Conservation Molecular ecology Behavioral ecology Urban ecology Physiological ecology Evolutionary ecology. The series comprise books and edited collections by international experts in their fields.

Takahisa Miyatake

Death Feigning

Mechanisms, Behavioral Ecology and Implications for Humans

Springer

Takahisa Miyatake
Faculty of Environment, Life, Natural
Science and Technology
Okayama University
Okayama, Japan

ISSN 2191-0707 ISSN 2191-0715 (electronic)
Ecological Research Monographs
ISBN 978-981-95-5318-1 ISBN 978-981-95-5319-8 (eBook)
https://doi.org/10.1007/978-981-95-5319-8

© The Editor(s) (if applicable) and The Author(s), under exclusive license to Springer Nature Singapore Pte Ltd. 2026

This work is subject to copyright. All rights are solely and exclusively licensed by the Publisher, whether the whole or part of the material is concerned, specifically the rights of reprinting, reuse of illustrations, recitation, broadcasting, reproduction on microfilms or in any other physical way, and transmission or information storage and retrieval, electronic adaptation, computer software, or by similar or dissimilar methodology now known or hereafter developed.
The use of general descriptive names, registered names, trademarks, service marks, etc. in this publication does not imply, even in the absence of a specific statement, that such names are exempt from the relevant protective laws and regulations and therefore free for general use.
The publisher, the authors and the editors are safe to assume that the advice and information in this book are believed to be true and accurate at the date of publication. Neither the publisher nor the authors or the editors give a warranty, expressed or implied, with respect to the material contained herein or for any errors or omissions that may have been made. The publisher remains neutral with regard to jurisdictional claims in published maps and institutional affiliations.

Cover Illustration: (Left) A beetle feigning death in front of a jumping spider. (Right) A tree frog playing dead.

This Springer imprint is published by the registered company Springer Nature Singapore Pte Ltd.
The registered company address is: 152 Beach Road, #21-01/04 Gateway East, Singapore 189721, Singapore

If disposing of this product, please recycle the paper.

Preface

Humans sometimes pretend to be dead—but why do we do this?

By the end of this book, I hope you will have found your own answer to this intriguing question. This book aims to provide a bird's-eye view of the emerging field of genomic behavioral ecology by linking ecology, physiology, and molecular biology through the lens of death-feigning behavior—and exploring its surprising connections to human relationships. Since the 2000s, research on death feigning or thanatosis—a peculiar yet widespread behavior observed across the animal kingdom—has gained traction as a compelling and rapidly advancing scientific field. Once considered anecdotal or peripheral, death feigning is now being systematically investigated using rigorous scientific methodologies, offering valuable insights into evolutionary biology, behavioral ecology, and animal psychology.

In Chap. 1, I share how I first became fascinated by the curious behavior of death feigning. I then explain, from an ecological viewpoint, how this behavior helps animals survive—not only by avoiding predators, but also in the struggles related to mating and social life. Predator-prey interactions represent a central theme in ecological science, attracting broad interdisciplinary interest. Among antipredatory strategies, death feigning stands out due to its remarkable taxonomic prevalence, experimental accessibility, and its implications across diverse fields—from neurobiology to human applications such as pest management and medical research. While numerous comprehensive texts explore antipredator behaviors broadly, none have yet provided an in-depth focus on death feigning. This book fills that gap by offering a thorough review of death feigning across multiple animal taxa, alongside a detailed account of the author's original experimental approaches and findings, thereby advancing our understanding of this intriguing survival strategy.

Chapter 2 reviews the history of research on death feigning and clears up some confusion around its definitions. I offer a clearer and more precise explanation of what we mean by immobility behavior. Who first identified death feigning, or thanatosis? This behavior has long been recognized as a form of antipredator strategy since the pioneering observations of Darwin and Fabre. However, recent research reveals that death feigning serves not only as an effective means of predator avoidance but also has broader adaptive significance. This chapter provides a concise overview of the scientific history surrounding experimental and observational studies on death feigning and explores the reasons for its widespread occurrence across animal taxa. Additionally, it clarifies and discusses the definition of death feigning.

While the subsequent chapters of this book will focus primarily on insects and their research history, this chapter highlights early visionary studies on death feigning in birds, as well as foundational research on vertebrates—particularly snakes and frogs—that contributed to our current understanding of this fascinating behavior. What internal factors regulate the duration of death feigning within an organism? This chapter presents experimental evidence using beetles that demonstrates the pivotal role of dopamine, a key biogenic amine, in modulating the length of death-feigning behavior. Additionally, it examines the impact of caffeine on death feigning and related behavioral responses. Furthermore, this chapter offers an overview of recent advances in neurophysiology pertaining to immobility and its underlying mechanisms.

Next, Chap. 3 lists the many animal species known to show death-feigning behavior, grouped according to their biological classifications. Death-feigning behavior has been documented across a diverse array of animal taxa. Table 3.2 summarizes the taxonomic groups in which death feigning, tonic immobility, or other related forms of immobility have been observed or reported. This chapter provides a comprehensive overview of immobility behaviors, categorizing them by taxonomic group and summarizing the existing research within each. The discussion begins with arthropods and proceeds to vertebrates. To date, death-feigning behavior has been reported in three animal phyla—Mollusca, Arthropoda, and Chordata—and has been observed in 12 distinct classes and 49 orders. For each order, this chapter presents the species in which death feigning has been reported, cites the relevant literature, specifies the terminology used to describe the behavior, and identifies the life stage during which the behavior was observed. Importantly, if readers observe death-feigning behavior in a taxonomic group not included in this list, such findings would represent a novel contribution to the field and merit publication in a peer-reviewed scientific journal.

Chapter 4 explores what triggers this behavior and what factors affect how often and how long animals play dead. Under what conditions do organisms exhibit death-feigning behavior? This chapter explores how various environmental and physiological factors influence the expression of death feigning, with a focus on experimental findings using beetles as model organisms. Specifically, it examines the effects of temperature, behavioral state, circadian rhythm, starvation, and the intensity of external stimuli. These studies provide insight into the context-dependent nature of this antipredator strategy and the environmental cues that modulate its occurrence. This chapter provides an overview of 24 factors associated with death-feigning behavior. It is dangerous for insects with death feigning to remain immobile all the time. They have to awaken at the right time from death feigning. In this chapter, I explain the strength of vibration causing arousal from the state of TI which was examined in strains artificially selected for longer duration of beetles derived from long strain, i.e., long sleeper. The results indicate a genetic basis for sensitivity to arousal from death feigning. Comparison of the timing of arousal from death feigning is also shown in the chapter. The other stimulations causing arousal from death feigning will be explained.

In Chap. 5, I describe the work I and my colleagues have done over the past 25 years studying beetles, revealing why death feigning is useful and how it works biologically. This chapter also reviews previous research on the genetic basis of death-feigning behavior, and presents the results of artificial selection experiments targeting the duration of death feigning in a beetle species. It discusses both the direct and correlated responses to selection, highlighting other traits that are genetically linked to the target trait—death-feigning duration. Furthermore, the chapter describes findings suggesting that the death-feigning syndrome is polygenically controlled, and emphasizes the value of reciprocal crossing experiments in elucidating the quantitative inheritance of behavioral traits. This chapter also presents a comparative analysis of gene expression profiles in strains of the beetle *Tribolium castaneum* that differ in their duration of death-feigning behavior. Beetles subjected to long-term artificial selection for either short or long death-feigning durations were analyzed using RNA sequencing. The results suggest that differences in the expression of certain enzymes involved in dopamine biosynthesis may influence the duration of this behavior. Additionally, the chapter discusses a comparative analysis of DNA variation between the selected strains. These findings indicate that multiple metabolic pathways and associated genes may play roles in the decision-making processes underlying death-feigning behavior.

Chapter 6 looks at the connection between death feigning and humans, including how it might relate to certain human conditions. I am confident that research on thanatosis in humans will continue to advance, particularly in connection with medical science. Two key avenues of development are especially promising. First, death-feigning behavior may involve molecular mechanisms—particularly those mediated by dopamine—that resemble those observed in human motor disorders such as Parkinson's disease. Elucidating these parallels could provide novel insights into the pathophysiology and potential treatment strategies for such conditions. Second, progress may come from investigations into the relationship between death feigning and human psychological phenomena, such as freezing responses observed in PTSD (Post-Traumatic Stress Disorder), hysteria, or during sexual assault. Understanding death feigning as a potential underlying mechanism in these contexts could have important implications for both clinical practice and legal interpretation. Understanding death feigning as a potential underlying mechanism in these contexts could have important clinical and legal implications. More broadly, some of the most compelling and complex questions about death-feigning behavior are becoming increasingly approachable through emerging research methods—including whole-genome sequencing and interdisciplinary approaches that integrate behavioral ecology, evolutionary biology, neuroscience, and medicine.

Finally, Chap. 7 discusses both the current challenges and future prospects for advancing research on death-feigning behavior, including the limitations inherent to laboratory-based studies. A key direction for future research will be to investigate the ecological relevance of death feigning—specifically, how it contributes to survival and reproductive fitness under natural field conditions—across a wide range of species. Questions that remain unanswered, suggestions for future research directions, and considerations of what this field may discover next. The book ends by asking you to reflect on a fundamental question: why do animals—including humans—pretend to be dead? And perhaps, in the insight offered by Darwin as early as 1883, we may glimpse the essence of death feigning—a discovery still unfolding. The animal kingdom is rich with examples of this mysterious behavior. You, too, may find yourself compelled to embark on the journey of exploration that lies ahead, into the fascinating world of death-feigning behavior. I hope.

Before delving into the main text, I would like to express my deepest gratitude to Professor Emeritus Yoh Iwasa (Kyushu University), who is also my primary doctoral advisor, and whose encouragement and support made the writing of this book possible. I am also profoundly thankful to Fumiko Yamaguchi of Springer for her invaluable editorial guidance throughout the publication process. Additionally, I am indebted to the many students and postgraduate researchers who have participated in the study of death-feigning behavior over the years, as well as to the numerous collaborators from other institutions who have generously shared their expertise and insight. To all of you, I extend my sincere and heartfelt thanks.

Okayama, Japan Takahisa Miyatake

Contents

Introduction 1

Abstract

This chapter provides an overview of each chapter on the subject of death feigning (thanatosis)—a tactic of remaining motionless as a strategy to avoid predation. Since Darwin (Mental evolution in animals. With a posthumous essay on instinct by Charles Darwin. London: Kegan Paul Trench & Co., 1883), death-feigning behavior has been documented across a wide range of animal taxa. Predator–prey interactions have been a central focus in the fields of animal behaviour and behavioural ecology for decades. While neurophysiological studies on immobility and psychological research on fear responses have a relatively long history, death feigning as an anti-predator strategy in an evolutionary ecological context remained largely understudied. Compared to other well-researched strategies such as crypsis, aposematism, active defence, and escape behaviour as ecological context, empirical studies on death feigning were scarce until the early 2000s. As a result, there was little data to demonstrate whether this behaviour actually contributes to survival. While working in the field of pest control, I became deeply fascinated when I observed that a species of weevil—considered a pest targeted for eradication—would become completely motionless in a distinctive posture when stimulated. That was the moment I first encountered death-feigning behaviour, which has become the focus of my research for over a quarter of a century.

Keywords

Darwin · Predator · Prey interaction · Animal defence · Avoiding attack · Sterile insect technique · Weevil

© The Author(s), under exclusive license to Springer Nature Singapore Pte Ltd. 2026

T. Miyatake, *Death Feigning*, Ecological Research Monographs,
https://doi.org/10.1007/978-981-95-5319-8_1

Predator–prey interactions have been a central focus in the fields of animal behaviour and behavioural ecology for decades (Edmunds 1974; Ruxton et al. 2004, 2018). Predation exerts a strong selective pressure on prey species (Lima and Dill 1990; Abrams et al. 2009; Caro 2005), driving the evolution of diverse antipredator behaviours that enhance survival during predator encounters (Ruxton et al. 2018). Animals employ a variety of strategies to avoid predation, including camouflage, warning coloration, mimicry (both concealed and aposematic), threat displays, and behavioural responses such as fight or flight (Edmunds 1974; Ruxton et al. 2018). Edmunds (1974) classified these strategies into primary and secondary defences. Primary defences are innate traits present regardless of predator presence—such as cryptic coloration, mimicry, and aposematism—whereas secondary defences are activated only upon detection or attack, including escape behaviours, aggression, or counterattacks. Among secondary defences, Edmunds (1974) briefly discussed death-feigning (thanatosis) in 'Defence in Animals', devoting only two pages to the topic, one of which was more than half occupied by illustrations.

Since Darwin (1883) first discussed the phenomenon of death feigning and Fabre (1910) provided detailed observational accounts, numerous studies have documented this behaviour across a range of taxa, including birds, snakes, amphibians, and insects (as detailed in Chap. 3). However, it was not until the study by Miyatake et al. (2004), using beetles and jumping spiders, that the adaptive value of death feigning was demonstrated statistically, based on robust empirical data. This work contributed to a renewed global interest in thanatosis research. Subsequent studies have confirmed that death-feigning behaviour is often directly linked to survival. For instance, death feigning has been observed not only in the context of predator avoidance but also in other ecological scenarios, such as in dragonflies avoiding conspecific harassment, spiders evading sexual cannibalism, and ants maintaining social cohesion.

As a preface to this account, it is pertinent to describe the origin of the author's interest in death-feigning behaviour. Although I have been fascinated by insects since childhood, my scientific interest in thanatosis did not begin until 1997. My first direct encounter with death-feigning behaviour occurred during research on the sweetpotato weevil (*Cylas formicarius*), a major pest of sweetpotato tubers. At the time, I was involved in a sterile insect technique (SIT) project aimed at eradicating this species from the southern islands of Japan (see Moriya and Miyatake 2001; Himuro et al. 2022; Ikegawa et al. 2022). As part of this effort, we conducted a series of mark—recapture experiments to estimate the dispersal capacity of the weevil—an essential parameter for the success of SIT programs (Miyatake et al. 1995, 1997, 2000). While rearing large numbers of *C. formicarius* individuals for release, I observed a striking behaviour: when gently prodded with a finger, the beetles would abruptly enter a rigid, motionless state, often in unusual postures (Fig. 1.1). This observation marked the beginning of my long-standing interest in death-feigning behaviour and its evolutionary and ecological significance.

Allow me to share a brief personal anecdote. At the time of my initial observations, I was formally engaged in pest management research on the sweetpotato weevil, *Cylas formicarius*. Given the applied nature of my professional responsibilities,

Fig. 1.1 The weevil *Cylas formicarius* feigns death in a unique posture

I initially hesitated to pursue a basic behavioural study on death feigning during official working hours. To resolve this internal conflict, I adopted a compromise: research on pest control was conducted during regular hours, while investigations into death-feigning behaviour were carried out after 5 p.m. and on weekends, when I was off duty.

Intrigued by the peculiar posture exhibited by the weevils during thanatosis, I began to make informal observations, quietly documenting the behaviour using laboratory-reared individuals. These initial explorations led me to question the adaptive significance of the pose, and in parallel, I began reviewing historical and contemporary literature on death-feigning behaviour across animal taxa. This period marked the informal yet formative beginning of what would become a long-term scientific inquiry into the phenomenon of thanatosis.

Over the subsequent 25 years—a full quarter of a century—I have continued to investigate the phenomenon of death-feigning behaviour. As my research progressed and I delved deeper into the relevant literature, it became increasingly apparent that, despite numerous descriptive reports across a wide range of taxa, there were remarkably few studies that provided empirical, quantitative evidence demonstrating that thanatosis directly contributes to individual survival. This gap in the literature underscored the need for experimental approaches to rigorously test the adaptive value of this intriguing behaviour.

From 1997 to 1998, I spent 1 year as a visiting researcher at University College London (UCL), where I investigated the mating and reproductive behaviour of the Mediterranean fruit fly (*Ceratitis capitata*) (Chapman et al. 1998; Miyatake et al. 1999). Upon returning to Japan, I resumed my work in Okinawa as a principal researcher on the control of the sweetpotato weevil (*Cylas formicarius*), while continuing observational studies on its death-feigning behaviour until the summer of 2000. In October 2000, I was appointed Associate Professor at Okayama University and relocated to the Japanese mainland. Having originally been born in Osaka and previously relocated to Okinawa as a university student, this marked my third major move within Japan. Due to the fact that *C. formicarius* is an invasive species in Japan—first recorded in 1919 and currently restricted to the southwestern islands—transporting live individuals to the mainland was prohibited under the Japanese Plant Protection Law. Consequently, I abandoned the use of *C. formicarius* as an experimental organism at Okayama University. Instead, I selected the red flour beetle (*Tribolium castaneum*) as an alternative model system to investigate the adaptive significance of death-feigning behaviour. Unlike agricultural research institutes, universities afford greater flexibility in the selection of research topics. Thus, in the autumn of 2000, I initiated a formal research programme on death-feigning behaviour, in collaboration with students in my laboratory, using beetles as the primary study system.

The results of our study were published in 2004, coinciding with the release of *Avoiding Attack* by Ruxton et al. (2004). Despite being published three decades after Edmunds' seminal work (*Defence in Animals*, 1974), Ruxton et al.'s volume devoted only a single section—Chap. 13.6—and two pages to the topic of death-feigning behaviour. In that chapter (titled "Tonic immobility, feigning death, pretending to be dead, animal hypnosis, thanatosis"), the authors explicitly noted that "feigning death has been neglected in recent years, resulting in a lack of data that meets modern standards of scientific rigour" (Ruxton et al. 2004). This observation strongly resonated with me. It confirmed my impression that the scientific study of death feigning had seen little methodological advancement and remained largely descriptive. I was both encouraged and motivated by this gap: if no one else was pursuing rigorous research on the subject, I resolved to do so myself. This conviction, fully formed in 2004, was built upon the curiosity first sparked in 1997, marking the beginning of a sustained and focused research programme. More than 25 years later, the study of death-feigning behaviour has expanded beyond traditional boundaries of behavioural ecology. As described in this book, it now intersects with a variety of research domains, including physiology, genetics, neurobiology, and even applied entomology.

In 2018, Ruxton et al. published the second edition of Avoiding Attack (Ruxton et al. 2018). To my great surprise and satisfaction, the revised edition included an entire chapter—Chap. 13—dedicated to thanatosis (death-feigning behaviour). Notably, this chapter cited more than ten publications in which my collaborators and I were involved. At the beginning of the chapter, the authors wrote:

> In the first edition we published, there was not enough material to write more than a few paragraphs on pretending to be dead. But fortunately, this situation has changed, and in the last 15 years there has been more purposeful and planned research into this phenomenon than at any time since Darwin (Ruxton et al. 2018).

This acknowledgement was deeply moving. While our contribution may be modest in the broader context of behavioural ecology, I believe it has played a small role in advancing scientific interest in death-feigning behaviour. The presence of our work in this chapter is a tangible indication that sustained empirical research over the past two decades has helped elevate thanatosis from a historically overlooked topic to one deserving of focused attention in mainstream literature. I am sincerely grateful to Professor Ruxton and his colleagues for recognising these contributions and for incorporating them into a major reference text on antipredator strategies. The inclusion of a dedicated chapter in the second edition serves as both validation and encouragement for continued research in this fascinating area of animal behaviour.

In recent years, death-feigning behaviour has attracted considerable interdisciplinary interest due to its taxonomic ubiquity, the ease with which it can be experimentally investigated, and its relevance across diverse research fields—including neurobiology, behavioural ecology, pest management, medicine, and even human psychology. While there are numerous comprehensive texts addressing antipredator behaviours in general, relatively few have focused specifically on death feigning. This volume aims to address that gap by providing an in-depth examination of death-feigning behaviour across animal taxa, alongside a detailed account of the author's own experimental work and findings.

One of the most fundamental behavioural distinctions in animals is the binary between movement and immobility. Beneath the seemingly simple observation that "an animal moves" lies an equally important counterpart: "an animal does not move." What factors determine whether an organism initiates or suppresses movement, especially in high-stakes ecological contexts? This question lies at the heart of the study of thanatosis.

Chapter 2 introduces definitions and traces the historical development of scientific inquiry into death-feigning behaviour. Chapter 3 reviews the occurrence of this behaviour across a broad range of taxa, drawing on both classical and recent literature. A comprehensive list of animal species in which death-feigning has been documented is provided in Table 3.1. Chapter 4 then explores the factors influencing the duration of thanatosis, which varies markedly depending on ecological, physiological, and genetic conditions. This chapter also outlines the trajectory of empirical research on this topic to date.

Chapter 5 presents a comprehensive account of the research on death-feigning behaviour that my collaborators and I have conducted over the past 25 years. This body of work reflects a sustained effort, supported by many dedicated colleagues, to address a deceptively simple yet fundamental question in behavioural biology: why do animals feign death? Throughout nearly a quarter of a century of inquiry, a central theme has consistently emerged—one that speaks to the core of behavioural decision-making in animals: the strategic dichotomy of movement versus immobility. Investigating the phenomenon of thanatosis has led us to uncover its deep associations with a wide array of biological processes and traits, including reproductive strategies, longevity, life-history trade-offs, dopaminergic signalling, genetic loci governing immobility duration, insect locomotion mechanics, and even clinical analogues such as Parkinson's disease.

In Chap. 6, I turn to the relevance of death-feigning behaviour beyond the animal kingdom, exploring its implications for human psychology and health. While thanatosis has traditionally been studied in non-human species, it is increasingly recognised as a phenomenon with deep connections to human mental states and trauma. In clinical and psychological contexts, this behaviour is typically referred to as tonic immobility, and is often observed in relation to extreme fear responses, particularly among survivors of sexual abuse or severe psychological stress. Research in these areas has evolved independently of zoological studies but converges on similar behavioural and neurophysiological mechanisms.

In this chapter, I offer a personal perspective on the potential links between death-feigning and human mental health conditions. While these reflections are necessarily tentative, they represent an effort to bridge gaps between behavioural ecology, neuroscience, and clinical psychology, and to consider the evolutionary roots and adaptive significance of such responses in humans.

The final chapter (Chap. 7) presents a synthesis of current knowledge—much of which remains fragmentary—and offers speculative insights into the evolutionary drivers of death-feigning behaviour. I also discuss unresolved questions and propose future directions for research in this still-developing field. Though we have made significant progress, death feigning remains one of the most intriguing and underexplored antipredator strategies in the animal kingdom, and its relevance across disciplines suggests it may yet yield many important discoveries.

The number of studies addressing death-feigning and tonic immobility has been steadily increasing in recent years. However, locating relevant references remains a significant challenge. This difficulty arises, in part, from the inconsistent terminology used to describe the behaviour—terms such as thanatosis, death feigning, tonic immobility, and animal hypnosis are not always employed uniformly across disciplines or taxonomic groups. For instance, some books on animals such as mantids or snakes may include descriptions of predator avoidance behaviours in sections titled "Behaviour" or "Anti-predator strategies." Yet these behaviours may be referenced only briefly, using non-standardised language—for example, "this animal sometimes pretends to be dead"—without explicit mention of thanatosis or tonic immobility. As a result, such entries are often not retrieved through keyword-based literature searches. Despite considerable effort to survey both physical and online sources using relevant terms, I acknowledge that there are likely references I was unable to locate. The limitations of individual searches are inevitable given the dispersed and occasionally ambiguous nature of the literature.

It is my hope that this book will serve as a foundational reference—or perhaps even a "bible"—for researchers around the world who are engaged in the study of death feigning, now and in the years to come.

An earlier version of some chapters in this book was originally published in Japanese in 2022 under the title "Shinda furi" de ikinobiru" by Iwanami Shoten, Publishers, Tokyo. These chapters are reproduced here with the permission of Iwanami Shoten.

References

Abrams MP, Carleton RN, Taylor S, Asmundson GJG (2009) Human tonic immobility: Measurement and correlates. Depress Anxiety 26:550–556. https://doi.org/10.1002/da.20462

Caro T (2005) Antipredator defenses in birds and mammals. University of Chicago Press

Chapman T, Miyatake T, Smith HK, Partridge L (1998) Interactions of mating, egg production and death rates in females of the Mediterranean fruit fly, *Ceratitis capitata*. Proc R Soc Series B 256:1879–1894

Darwin CR (1883) A posthumous essay on instinct. In: Romanes GJ (ed) Mental evolution in animals. With a posthumous essay on instinct by Charles Darwin. Kegan Paul Trench & Co., London, pp 188–189, 196, 198–199, 301–302, 355–384

Edmunds M (1974) Defence in animals. Longman, London

Fabre JH (1910) Souvenirs entomologiques. Delgrave Paris, Paris

Himuro C, Kohama T, Matsuyama T, Sadoyama Y, Kawamura F, Honma A, Ikegawa Y, Haraguchi D (2022) First case of successful eradication of the sweet potato weevil, *Cylas formicarius* (Fabricius), using the sterile insect technique. PLoS One. https://doi.org/10.1371/journal.pone.0267728

Ikegawa Y, Kawmura F, Sadoyama Y, Kinjo K, Haraguchi D, Honma A, Himuro C, Matsuyama T (2022) Eradication of sweetpotato weevil, *Cylas formicarius*, from Tsuken Island, Okinawa, Japan, under transient invasion of males. J Appl Entomol 146:850–859

Lima SL, Dill LM (1990) Behavioral decisions made under the risk of predation: a review and prospectus. Can J Zool 68:619–640

Miyatake T, Kawasaki K, Kohama T, Moriya S, Shimoji Y (1995) Dispersal of male sweetpotato weevils (Coleoptera: Curculionidae) in fields with or without sweet potato plants. Environ Entomol 24:1167–1174

Miyatake T, Moriya S, Kohama T, Shimoji Y (1997) Dispersal potential of male sweetpotato weevils (Coleoptera: Brentidae) over land and water. Environ Entomol 26:272–276

Miyatake T, Chapman T, Partridge L (1999) Mating-induced inhibition of remating in female Mediterranean fruit flies *Ceratitis capitata*. J Insect Physiol 45:1021–1028

Miyatake T, Kohama T, Shimoji Y, Kawasaki K, Moriya S, Kishita M, Yamamura K (2000) Dispersal of released male sweetpotato weevil, *Cylas formicarius* (Coleoptera: Brentidae) in different seasons. Appl Entomol Zool 35:441–449

Miyatake T, Katayama K, Takeda Y, Nakashima A, Sugita A, Mizumoto M (2004) Is death-feigning adaptive? Heritable variation in fitness difference of death-feigning behaviour. Proc R Soc Lond Ser B 271:2293–2296

Moriya S, Miyatake T (2001) Eradication programs of two sweetpotato pests, *Cylas formicarius* and *Euscepes postfasciatus*, in Japan with special reference to their dispersal ability. JARQ 35:227–234

Ruxton GD, Sherratt TN, Speed MP (2004) Avoiding attack. Oxford University Press

Ruxton GD, Allen WL, Sherratt TN, Speed MP (2018) Avoiding attack: the evolutionary ecology of crypsis, aposematism, and mimicry. Oxford University Press, Oxford

History and Definition of Death Feigning 2

Abstract

This chapter begins with an overview of references to death-feigning behavior in media—ranging from literature and performing arts to politics. It then introduces Darwin's thoughts on the subject and Fabre's pioneering experiments using beetles, with a historical overview of research on death feigning. Next, I examine the definition of death feigning, which has long been a source of confusion. Based on this background, I propose a novel and clear definition of death feigning. In this new framework, a distinction is made between various types of immobility observed across taxa—specifically between "freezing," "tonic immobility," and "death feigning." A table is provided to clearly differentiate these terms, along with an explanation of the contexts in which each is used. I then present five leading hypotheses that have been proposed to explain the adaptive significance of death feigning. In addition, I discuss studies that have addressed death feigning in the context of evolutionary ecology, particularly from the perspective of fitness. Finally, I explain the background that led me to this proposal. I explore when, in what situations, and under what conditions insects engage in death feigning—addressing the classic "5W1H" (who, what, when, where, why, and how) of this intriguing behavior.

Keywords

Fabre · Darwin · Definition of death feigning · Historical view · Adaptive significance · Active mode · Inactive mode

© The Author(s), under exclusive license to Springer Nature Singapore Pte Ltd. 2026

T. Miyatake, *Death Feigning*, Ecological Research Monographs,
https://doi.org/10.1007/978-981-95-5319-8_2

2.1 Darwin and Fabre

The concept of death feigning, or thanatosis, is not only observed in animal behavior but also appears recurrently in human cultural and historical contexts. In the well-known Chinese historical chronicle "The Romance of the Three Kingdoms," there is a passage describing a general who reportedly survived by feigning death (Yoshikawa 1989). Such instances likely occurred throughout history and across cultures, suggesting that death feigning may have conferred survival advantages even in human conflict scenarios. A notable political example in Japan is the so-called "dead pretender dissolution" (shinda furi kaisan) of the House of Representatives on June 2, 1986, in which lawmakers abruptly and unexpectedly dissolved the legislative body. Public discourse also frequently addresses the question of whether feigning death is an effective strategy when encountering wild animals such as bears. The motif further appears in domestic and social contexts, where anecdotal references include individuals pretending to be dead to avoid confrontation with their spouses, or vice versa. Most recently, the term has even entered popular culture: in June 2025, the idol group Keyakizaka46 released a song titled Shindafuri ("Feigning Death"), highlighting the term's familiarity in everyday language and media (Keyakizaka46 Official YouTube Channel 2025). These diverse examples illustrate how the notion of death feigning extends beyond biological behavior, permeating various aspects of human society, history, and cultural expression.

Turning to the scientific domain, one of the earliest recorded observations of death-feigning behavior in animals was made by Charles Darwin, best known for his theory of evolution. In an essay published in 1883, Darwin described instances of immobility and death feigning as instinctive behaviors exhibited by animals as a means of avoiding predation (Darwin 1883). Following Darwin's observations, numerous scientists dedicated considerable effort to studying the behavior of living organisms, particularly insects. Among them, Jean-Henri Fabre made substantial contributions. Known for his meticulous naturalistic observations, Fabre investigated death-feigning behavior primarily in beetles, often conducting experiments in domestic settings. In his widely recognized work Souvenirs Entomologiques, he devoted a chapter to this phenomenon, formulating hypotheses and reporting his experimental findings (Fabre 1910). From a scientific perspective, one of the earliest documented reports of death-feigning behavior in animals was made by Charles Darwin, who is renowned for formulating the theory of evolution. In an essay published in 1883, Darwin discussed instances of immobility and tonic immobility (commonly referred to as death feigning) as instinctive behaviors that serve as a defense mechanism against natural enemies (Darwin 1883).

In the years following Darwin's work, numerous scientists devoted substantial effort to observing animal behavior, particularly among insects. One of the most prominent figures in this regard was Jean-Henri Fabre, who conducted extensive observations of death-feigning behavior, primarily in beetles, often within domestic settings. In his influential work Souvenirs Entomologiques, Fabre dedicated an entire chapter to the subject, presenting a series of hypotheses and experiments related to tonic immobility (Fabre 1910). Despite his careful and detailed observations, Fabre was ultimately unable to determine when, exactly, death feigning occurred in beetles, or what the substantial variation in its duration might signify. He was particularly puzzled by the apparent inconsistency in behavioral patterns and questioned the adaptive function of the behavior. As he wrote: "I would like to extract from this seemingly chaotic duration of death feigning some certain law of behaviour" (Fabre 1910). His reflections suggest a degree of skepticism regarding whether this behavior effectively contributes to predator avoidance.

During Fabre's time, technologies such as precision weighing instruments, temperature-controlled environments, and advanced genetic analysis tools were not yet available. One cannot help but wonder how much further his discoveries might have progressed had he had access to the experimental precision and technological resources available in modern scientific research. Despite continued interest, the question of whether death-feigning behavior provides a survival advantage—particularly from the perspective of evolutionary ecology—has yet to be conclusively answered using robust quantitative data. To date, no study has provided definitive empirical evidence to resolve this issue. In contrast, extensive research has been conducted on the natural history and physiological mechanisms underlying death-feigning behavior, resulting in a substantial accumulation of descriptive and mechanistic knowledge in this area.

2.1.1 Natural History and Physiology

According to Wikipedia (https://en.wikipedia.org/wiki/Apparent_death), the earliest documented reference to death-feigning behavior dates back to Athanasius Kircher in 1646, who described the phenomenon as a form of "animal hypnosis" (cited in Gilman et al. 1950). In the field of insect physiology, the behavior has long been studied under the terms thanatosis and animal hypnosis, with numerous early reports detailing its occurrence across various taxa (e.g., Robertson 1904; Holmes 1906). In some species, death feigning is accompanied not only by immobility but also by physiological changes such as reductions in body temperature, heart rate, and respiratory rate. In addition, certain animals emit noxious odors or exude bodily fluids, potentially mimicking signs of disease to deter predators. The behavior often referred to as immobility—a state in which the organism ceases movement in response to external stimuli—was historically equated with death feigning. A more formal conceptualization appeared in Edmunds' seminal work Defence in Animals: A Survey of Anti-Predator Defences (Edmunds 1974), which was among the first to classify and describe death-feigning behavior systematically as a specific antipredator strategy. However, it is noteworthy that only two pages of the volume were devoted to this behavior, indicating its relatively marginal status in early comparative studies of animal defense mechanisms.

Animals exhibiting mimicry often enter a state of immobility while adopting a fixed posture. The specific posture varies across species, but it is typically unnatural and differs markedly from the animal's usual resting or locomotory positions. When an external force attempts to alter this posture, the animal typically responds immediately by attempting to restore its original position—this response is referred to as the resistance reflex. In contrast, during death-feigning behavior (thanatosis), this resistance reflex becomes transiently impaired, allowing the animal to maintain an unnatural posture without active correction. This phenomenon is known as catalepsy, and it is considered a common characteristic of animals undergoing thanatosis. The duration of death feigning generally ranges from several minutes to several tens of minutes, although considerable interspecific variation exists. In many species, the return to normal activity occurs abruptly. Mechanical stimulation—such as prodding with a stick—often terminates the immobile state and elicits arousal. Interestingly, in some beetle species, such stimulation can paradoxically prolong the death-feigning response rather than shorten it.

Nishino and Sakai (1996) investigated thanatosis in the cricket *Gryllus bimaculatus*. During the death-feigning state, the animals exhibited a significant reduction in respiratory rate and responsiveness to external stimuli. The immobile posture typically persisted for 2–4 minutes, though in some cases it lasted for more than 20 minutes, and termination of the state occurred abruptly. Catalepsy was observed during thanatosis when a leg was passively extended; the leg remained in the imposed position without resistance. The flexion reflex was found to require input from campaniform sensilla and femoral chordotonal organs (FCOs) in the forelegs, as well as spine-like sensilla on the pronotum. However, only the FCOs were implicated in initiating and maintaining the immobile state. Centrally, brain activity was shown to be essential for the maintenance of thanatosis. Subsequent work by Nishino et al. (1999) used this behavioral paradigm to functionally differentiate subgroups of identified neurons within the FCO system. They found that the muscular state during thanatosis differed from that in normal resting animals: muscles exhibited increased rigidity, enabling the maintenance of a fixed posture for extended periods.

Furthermore, Nishino et al. (1999) proposed that behavioral quiescence and thanatosis, though outwardly similar as forms of immobility, are distinct states that are differentially regulated by the central nervous system. In a later study, Nishino (2004) examined the mechanical and physiological properties of the metathoracic tibial flexor muscle and its posture-related activation. He reported that the motor neuronal pool is actively suppressed during thanatosis, and that flexor muscle rigidity is maintained by low-frequency discharge from slow excitatory motor neurons in combination with concurrent inhibition of inhibitory motor neurons.

2.1.2 The Terminological Problem of Death Feigning: A Historical Overview

To avoid terminological ambiguity, we consistently use the term death feigning, which we define as "a state of immobility accompanied by a distinctive posture, elicited in response to external stimuli." However, a variety of synonymous terms have been used across scientific literature to describe this behavior. These include playing dead, pretending to be dead, and apparent death. In North America, the behavior is colloquially referred to as playing possum, named after the opossum (*Didelphis virginiana*), which is well known for this antipredator strategy—often accompanied by the emission of foul odors that mimic the scent of decay.

In physiological contexts, the behavior is often referred to as thanatosis, derived from the Greek word Thanatos (death). Additionally, related terms such as animal hypnosis and catalepsy have been used, particularly in earlier literature, to describe states of immobility or muscular rigidity in animals. However, these terms are sometimes difficult to differentiate clearly from death feigning and may refer to overlapping but not identical phenomena.

Rogers and Simpson (2014) wrote that the term *thanatosis* is derived from the Greek word for death and refers to a distinctive behavioral state that is known by various names, including "death feigning," "playing possum" (in reference to one of its most iconic exponents), "catatonia," and the more colloquial "animal hypnosis." Among these, the term *tonic immobility* is most widely adopted in scientific discourse due to its relative neutrality. Thanatosis is characterized by several distinct physiological and behavioral features. These include catalepsy—an immobile yet rigid posture maintained through sustained tonic muscle contraction; "waxy flexibility," wherein an externally manipulated limb retains its new position for an extended duration; and a marked unresponsiveness to external stimuli despite apparent awareness of the surrounding environment. The onset of thanatosis is typically rapid and can last from several seconds to multiple hours. Termination is usually abrupt, with the immediate reactivation of the righting reflex, enabling the animal to resume full motor function without observable impairment.

In other words, the terminology used to describe the phenomenon in which animals suddenly become immobile varies depending on the research discipline addressing it. Physiologists who study underlying mechanisms tend to use terms such as tonic immobility or thanatosis. Behavioral psychologists may refer to the phenomenon as animal hypnosis or catalepsy—the latter describing a state in which an animal maintains a posture imposed upon it, seemingly against its will. In contrast, ethologists and animal behaviorists commonly use terms such as freezing or death feigning (also referred to as playing dead). For example, Acheampong and Mitchell (1997) described freezing behavior in beetles as a form of quiescence. Additionally, some behavioral researchers in the United States have recently advocated for the term post-contact immobility to more precisely describe immobility following predator contact.

This wide range of terminology—despite describing outwardly similar behaviors—reflects a lack of consensus across disciplines. The absence of a unified vocabulary, along with the independent definitions proposed by different researchers, has made the classification of immobile behavior particularly complex. Further complicating this issue is the difficulty in distinguishing whether the immobility is actively controlled (i.e., a form of mimicry or strategic behavior) or passively induced (e.g., resulting from overwhelming external force or stress).

This raises a fundamental question: who can truly be defined as a death feigner? If terms such as playing dead or death feigning are to be defined strictly, it would imply that the animal performing the behavior, as well as the observer (e.g., predator or conspecific), possesses an awareness of the concept of death and interprets the act accordingly. However, it is highly challenging—if not impossible—to determine through behavioral observation alone whether animals can reliably distinguish between actual death and mere immobility, or whether they respond based on such a distinction.

To date, there appear to be no studies that explicitly address this cognitive distinction or provide a strict operational definition of pretending to be dead that accounts for the perceptual and interpretive capacities of both the actor and the observer. This highlights the need for a definition that is not based on the reaction of the recipient (e.g., predator behavior), but rather on the observable characteristics of the actor's behavior itself. Such an approach would allow for a more objective, functional classification of death-feigning behavior across taxa.

This raises a critical conceptual question: what qualifies an organism as a death feigner? If we are to define terms such as death feigning or playing dead in a strict sense, it would logically require that the animal performing the behavior, as well as the observer (e.g., predator or conspecific), possesses some awareness of the concept of death and interprets the behavior in that context.

However, in practice, it is extremely difficult to determine whether nonhuman animals can cognitively distinguish between actual death and immobility, or whether their behavioral responses are based on such a distinction. Such assessments cannot be reliably made through behavioral observation alone, and, to the best of our knowledge, no existing studies have empirically addressed this cognitive differentiation or proposed a rigorous definition of what it means for an animal to "pretend to be dead."

Therefore, there is a clear need to redefine death-feigning behavior based not on the reaction of the recipient (e.g., the predator's interpretation), but on the observable characteristics and intentional features of the actor's behavior itself. A definition grounded in the animal's behavioral expression—rather than in the inferred perception of an observer—would provide a more objective and operational framework for comparative and experimental studies.

Throughout this work, I consistently define death feigning as "a distinctive posture adopted by an organism in which it ceases movement in response to an external stimulus." The emphasis in this definition lies in the phrase a distinctive posture. This criterion is critical to distinguishing death feigning from other forms of immobility. Specifically, the mere cessation of movement following an external stimulus—without a change in posture—is more appropriately referred to as freezing. In freezing, the animal remains immobile while maintaining its preexisting or natural position, rather than adopting an unusual or defensive posture. This distinction is supported by recent literature (see Asakura et al. 2022) and is upheld throughout this book when referring to death-feigning behavior.

The term tonic immobility has long been used in the fields of physiology and medicine, and is particularly prevalent in descriptions of immobility in birds. Another commonly used term is freezing, which refers to a behavioral response in which an animal ceases movement upon detecting an external presence (Sakai 2021a, b; Asakura et al. 2022).

As noted earlier, similar immobility behavior has been described in Colorado potato beetles and referred to as quiescence (Acheampong and Mitchell 1997). However, based on behavioral characteristics, we consider there to be no clear or functional distinction between freezing and quiescence. Both terms appear to describe the same phenomenon—passive immobility without the adoption of a distinctive or defensive posture.

Following the early reflections by Darwin and Fabre, research on death-feigning behavior was primarily advanced by insect physiologists. These studies focused on the physiological mechanisms underlying immobility and typically employed terminology such as tonic immobility and catalepsy, rather than the term death feigning.

This preference in terminology reflects a disciplinary distinction. The phrase feigning death carries a secular and somewhat anthropomorphic connotation, implying a conscious or intentional act of deception. In contrast, terms like tonic immobility and catalepsy are rooted in physiological and neurobehavioral frameworks, emphasizing involuntary motor inhibition rather than deliberate behavioral mimicry.

Physiologists primarily focused on the neural and muscular mechanisms responsible for movement in the peripheral parts of the body. This emphasis was, in part, due to the limitations in accessing and studying the neural correlates of consciousness—particularly in insects, where direct investigation of brain function at the cognitive level was not feasible with the available techniques. As a result, research concentrated on observable motor phenomena rather than attempting to infer intentionality or awareness.

Behavioral psychologists have distinguished the freezing response—a state in which birds and other animals become immobile when experiencing fear—from death-feigning behavior, treating them as distinct forms of immobility. The development of various classifications of immobility within behavioral psychology has likely contributed to the proliferation of overlapping and sometimes ambiguous terminology.

A central question in this field concerns whether such immobility—be it freezing or death feigning—is a conscious, intentional act. Specifically, does the prey animal deliberately remain motionless because it has learned or evolved to recognize that predators may cease their attack in response to apparent death? Or, alternatively, is the immobility an involuntary physiological reaction that happens to confer a survival advantage by reducing detection?

This distinction touches on a deeper and unresolved issue concerning the role of cognitive processing, awareness, and evolutionary function in antipredator behavior. Understanding whether these responses are under voluntary control or are purely automatic has significant implications for how we interpret animal behavior and its underlying neural mechanisms.

A distinction is often drawn between freezing and death-feigning behaviors. A widely accepted view among behavioral researchers is that freezing represents an initial stage in a sequence of antipredator responses, while death feigning typically occurs later in the sequence—as a so-called "last resort" defense strategy (see Sakai 2021a, b). However, our recent findings suggest that this sequential framework does not always apply. In some cases, the two behaviors do not follow a fixed progression and may occur independently of each other (Asakura et al. 2022).

2.1.3 Proposed Conceptual Framework: Distinguishing "Freeze," "Immobility After Contact," and "Death Feigning"

Sakai (2021a, b) previously discussed the distinction between freezing and tonic immobility as forms of immobility behavior. However, the study did not provide clear criteria for differentiating between the two. In addition, the term *animal hypnosis*, another expression referring to immobility, was cited from Klemm (1977) and applied to humans. Since then, new terms such as *post-contact immobility* and simply *immobility* have also come into use. Nevertheless, there has been no attempt to clearly define and distinguish these terms based on specific criteria.

To clarify the confusion surrounding the terminology used to describe immobility behaviors in prey species, I propose a conceptual distinction among three commonly used terms: freeze, immobility after contact, and death feigning (See Table 2.1). While these behaviors may appear superficially similar in that they

Table 2.1 Anti-predator Immobility Terminology—Classification by Posture and Stimulus Timing

Classification criteria	Natural posture	Distinctive posture
Pre-contact	Freeze	(Not typically observed)
	Immediate immobility upon detecting threat	
	No change in posture	
	Common in behavioral psychology (e.g., Sakai 2021a, b)	
Post-contact	Post-contact immobility/Tonic immobility	Death feigning/Thanatosis
	Immobility after predator touch or grasp	Immobility with unusual or rigid posture
	No distinct posture	Often includes muscle stiffness or odor
	Used in physiology, medicine, psychology	Used in ethology, behavioral ecology

Freeze: A primary anti-predator response aimed at avoiding detection by visually oriented predators. The animal abruptly ceases movement but maintains its natural posture without adopting any abnormal pose

Post-contact immobility/Tonic immobility: A state of immobility that occurs following direct physical contact with a predator. The posture remains unaltered, but the behavior is believed to involve neural and reflex suppression mechanisms, often studied in neurophysiology and behavioral psychology

Death feigning/Thanatosis: A defensive behavior in which the animal adopts an abnormal or rigid posture after predator contact, mimicking death. This behaviour is thought to reduce the predator's motivation to continue its attack, functioning as a form of deceptive antipredator strategy

involve a cessation of movement, they differ in context, posture, triggering stimuli, and potential functional interpretations.

2.1.3.1 Freeze

Freeze refers to an immediate cessation of voluntary movement in response to the detection of a potential threat, often at a distance. The animal typically retains its natural or neutral posture, and the behavior may be interrupted quickly upon further environmental change. Freeze behavior is considered an early-stage antipredator response, intended to reduce detection by predators that rely on visual cues (Sakai 2021a, b; Asakura et al. 2022).

2.1.3.2 Immobility After Contact

Immobility after contact (also called post-contact immobility) describes a state in which an animal becomes motionless following physical contact or close proximity with a predator. This term emphasizes the temporal context of the behavior—occurring after a predator makes physical contact or engages closely. Unlike simple freezing, this immobility may be more prolonged and is often associated with a sudden transition into a defensive state. Some researchers propose this as a distinct category that overlaps with, but is not identical to, death feigning.

2.1.3.3 Death Feigning

Death feigning (also known as thanatosis or playing dead) involves not only immobility but also the adoption of a specific, often abnormal or unnatural posture. It typically occurs under high-threat conditions and is thought to function as a form of defensive mimicry—imitating the appearance of death to deter predator interest. The behavior is often accompanied by additional cues such as odor emission or muscle rigidity, and may involve suppression of reflex responses (Fabre 1910; Nishino and Sakai 1996; Asakura et al. 2022).

This new arrangement aims to facilitate more precise communication among researchers and promote cross-disciplinary understanding. By distinguishing these behaviors based on triggering context and observable features, we can better interpret their adaptive significance and underlying mechanisms. New arrangement: words "freeze," "immobility after contact," "death feigning."

I am aware that there is ongoing debate about referring to behaviors such as pill bugs curling into a ball and becoming immobile, or grasshoppers extending their legs and freezing, as thanatosis (see Humphreys and Ruxton 2018). However, I repeat: I define death feigning (a synonym of thanatosis) as "*suddenly becoming immobile with a distinctive posture in response to an external physical stimulus.*" Without such a definition, I fear discussions of immobility behavior will be dominated by unproductive debates over terminology.

I have used the term freeze to describe the behaviour of becoming immobile in response to stimuli by an enemy, which does not involve a distinctive pause in doing so, but simply a halt in movement. And two words have been used to describe the unique posed immobility: death feigning and tonic immobility. However, as described in previous studies (Humphreys and Ruxton 2018; Sakai 2021a, b), immobile behaviour involves a series of occasions during which the animal perceives predator danger and expresses immobile behaviour with different meanings before actually being attacked. I therefore propose anew that a series of words be organized and used as follows.

2.1.4 Stage 1: Detecting the Presence of the Enemy

When an animal perceives a potential predator, it may cease movement without altering its existing posture, simply interrupting an ongoing activity such as walking or running. This form of immobility—characterized by a lack of postural change—has been described in the literature using a variety of terms, including freeze, immobility, quiescence, and tonic immobility.

While freezing has historically been the most commonly used term for this behavior (e.g., Sakai 2021a, b), other terms have also been applied to describe what appears to be the same behavioral phenomenon. Below, I highlight key examples that illustrate this terminological inconsistency:

(a) The term immobility was used by Katsuki et al. (2024) to describe predator-avoidance behavior in the horned flower beetle (*Gnatocerus cornutus*), where individuals stopped moving in response to a threat. This usage is effectively equivalent to freezing. Notably, in the same paper, behaviors previously described as tonic immobility in the literature were also subsumed under the general label of immobility, further blurring the distinction between terms.
(b) In a study on the Colorado potato beetle (*Leptinotarsa decemlineata*), Acheampong and Mitchell (1997) used the term quiescence to describe a similar form of stimulus-induced immobility in adult beetles. Here too, the behavior is functionally indistinguishable from what is typically described as freezing.
(c) The term tonic immobility has introduced considerable confusion in the literature, as it has been applied to describe both immobility without a change in posture and immobility with the adoption of a distinctive, defensive posture. For example, in Miyatake et al. (2009) and Nakayama and Miyatake (2009), tonic immobility is used in both senses, making it difficult to draw a clear behavioral or mechanistic distinction.

These examples underscore the pressing need to clarify and standardize the terminology used to describe immobility responses to predation threat, particularly with respect to whether or not a change in posture occurs.

2.1.5 Stage 2: Immobility Triggered After Being Attacked by the Enemy

In the scientific literature, two principal terms—death feigning and thanatosis—are commonly employed to describe the immobilization of prey animals immediately following physical contact with a predator (Humphreys and Ruxton 2018). Additionally, the term tonic immobility is frequently used, particularly within the domains of neurophysiology, behavioral psychology, and medicine, where it typically refers to a sudden, temporary state of motor inhibition observed in birds and other vertebrates. Notably, this term is also occasionally applied to similar behaviors in insects, leading to interdisciplinary overlap and potential terminological ambiguity. The concurrent use of these overlapping terms to describe behaviorally analogous, yet potentially mechanistically distinct, phenomena across different scientific fields has likely impeded the development of a unified and operational definition for immobility-based antipredator strategies.

The term post-contact immobility—often used interchangeably with tonic immobility—has been broadly applied to encompass all forms of immobilization that occur in response to direct external stimuli, such as physical contact with a predator. In contrast, when such immobility is accompanied by the adoption of a distinctive and often unnatural posture, the behavior is more precisely referred to as death feigning or thanatosis. These latter two terms are generally regarded as synonymous and provide a more specific and less ambiguous classification of this particular behavioral response.

To promote terminological precision across disciplines, it is advisable to restrict the use of death feigning or thanatosis to instances of immobility that are accompanied by distinctive and atypical postural changes. In contrast, post-contact immobility or tonic immobility may serve as broader umbrella terms encompassing all immobilization responses elicited by direct predator interaction. Establishing such a distinction is likely to facilitate conceptual consistency and minimize ambiguity in the cross-disciplinary investigation of antipredator behaviors.

If post-contact immobility is defined strictly within the context of predator–prey interactions, then an immobile animal that is perceived by the predator as dead—and therefore no longer as prey—should be classified as exhibiting death feigning or thanatosis. In support of this conceptual framework, the five hypotheses previously proposed by Miyatake et al. (2009) may serve as a useful basis for organizing and interpreting such behaviors. These hypotheses are cited below.

1. The "Avoiding Dead Prey" Hypothesis: If a predator is capable of discerning that a prey item is dead, it may avoid consuming it due to the potential risk of disease transmission. In this narrow sense, the term death feigning should be applied only when the predator possesses the ability to distinguish between live and dead prey.
2. The "Concealment and/or Background Matching" Hypothesis: When an insect falls from a perch, it often enters a state of tonic immobility upon landing. If the insect's dark body coloration blends with the substrate, the predator may lose visual contact with the prey (Ohno and Miyatake 2007). In this context, tonic immobility contributes to concealment through background matching.
3. The "Physical Defence" Hypothesis: A grasshopper exhibiting tonic immobility may assume a distinctive posture that increases its apparent body size by extending its limbs and other body parts. This expansion of functional body dimensions

can impede ingestion by predators such as frogs. In this context, tonic immobility serves as a form of physical defense (Honma et al. 2006).
4. The "Loss of Predator's Interest" Hypothesis: Tonic immobility may serve as an adaptive defense against predators that are stimulated primarily by prey movement (Miyatake et al. 2004). When the prey becomes immobile, the predator may lose interest and abandon the attack, opting instead to search for alternative targets.
5. The "Chemical Alarm Signalling" Hypothesis: The posture adopted by prey during tonic immobility may serve as a signal of unpalatability to potential predators (Ruxton 2006). Predators can learn to avoid consuming organisms that possess strong chemical defenses, and immobility—particularly when coupled with specific postures—may function as an honest cue indicating such defenses.

In summary, to enhance reader comprehension, this book adopts the following terminology to describe antipredator immobility behaviors, classified according to the stage of predator–prey interaction in which they occur:

1. *Freeze* (synonyms: *immobility*, *quiescence*): Refers to immobility that occurs at *Stage 1*, immediately upon detection of a predator, before any physical contact is made. This behavior involves the cessation of movement without a change in posture and is thought to reduce the likelihood of detection by visually oriented predators.
2. *Post-contact immobility* (synonym: *tonic immobility*): Refers to immobility that is triggered at *Stage 2*, following direct contact or an attack by a predator. Within this category, the book distinguishes between two subtypes:

   - *Post-contact immobility without postural change* (sometimes termed *quiescence*): A passive immobility response that lacks any conspicuous change in body posture.
   - *Death feigning* (synonym: *thanatosis*): A subtype of post-contact immobility in which the animal adopts a distinctive or unnatural posture. This behavior is redefined here to include the possibility that such postures cause the predator to perceive the prey as inedible or non-prey, thereby inhibiting further attack.

2.1.6 Adaptive Value of Death-Feigning Behavior in Antipredator Defense

This section explores the question of whether feigning death can truly help an individual evade attack or danger. The idea that immobility might confer a survival advantage has intrigued not only scientists but also the general public.

In 1883, Charles Darwin briefly addressed this phenomenon in his short report An Essay on Instincts. He noted: "Animals feign death. This unknown state of affairs for the organism seemed to me to be a remarkable instinct. (...) It is certain that the paralyzing effects of fainting or excessive fear can be mistaken for a simulation of death." To the best of my knowledge, this was the first scientific account to consider death-feigning behavior in animals, marking an important early step in the study of its adaptive significance.

Later, in 1900, Jean-Henri Casimir Fabre reported detailed observations of death-feigning behavior in insects in a chapter titled Death Feigning within his work Fabre's Insect Chronicle (Complete Fabre Insect Chronicle 7 1993). In a series of home-based experiments, Fabre placed beetles on his desk, stimulated them to induce immobility, and carefully observed the timing of their recovery. He was particularly intrigued by the duration of the death-feigning state. Despite conducting multiple trials under varying conditions—including different days and ambient temperatures—he was unable to determine the precise factors influencing the length of immobility.

In another experiment, Fabre induced a small-bodied beetle to feign death on a table and then introduced a larger beetle into the same space. As the larger beetle approached, the smaller one abruptly ended its immobility and fled. This observation suggested that the decision to terminate death feigning may be context-dependent and responsive to perceived threat levels. At one point, Fabre remarked with a tone of despondency, "This is the time to feign death." As his investigations continued, he grew increasingly uncertain about the adaptive value of death-feigning behavior in beetles. Despite conducting numerous additional experiments, he found himself unable to reach a clear conclusion. In a moment of candid self-reflection, he asked, "Can I get any satisfactory conclusions out of this jumble of muddled things? I hope so." With this unresolved question, he brought the chapter to a close.

Both Darwin and Fabre attempted to interpret death feigning either as a behavioral strategy to avoid predation or as a reaction triggered by extreme fear in the face of a predator. Subsequent researchers have generally accepted this behavior as an antipredator strategy, equating death feigning with evolved predator avoidance. But does feigning death truly reduce the likelihood of being consumed by a predator? And if so, can it be considered an evolved behavior precisely because it provides a selective advantage? This question has remained central to understanding the adaptive significance of death-feigning behavior. There have been three studies that have looked into solving this puzzle. To date, three key studies have directly addressed this evolutionary puzzle.

Three empirical studies have directly investigated whether death-feigning behavior provides a survival advantage against predators.

The first was conducted in North Dakota in 1975 by Sargeant and Eberhardt (1975). In this study, 29 of 50 ducks that exhibited death-feigning behavior successfully escaped predation by red foxes, suggesting that immobility may reduce the likelihood of capture.

The second study, carried out in Pennsylvania (Thompson et al. 1981), exposed 32 quails to attacks by Aedes aegypti. Among 16 recorded attacks, 14 were directed at quails that were actively moving, while quails that remained motionless were never attacked, indicating that immobility may inhibit predator targeting.

The third study, by Arduino and Gould (1984), found that chickens exhibited significantly longer durations of tonic immobility in the presence of predatory birds (e.g., hawks) or images mimicking avian eyes, compared to control conditions. This suggests that the perceived threat of predation enhances the expression of death-feigning behavior.

Though conducted under controlled conditions resembling the wild, all three studies suggest that feigning death can be an effective means of predator avoidance.

In the twentieth century, only three studies—those described above—explicitly addressed whether death-feigning behavior is adaptive from the perspective of fitness and evolutionary ecology. That is, they considered whether individuals possessing genotypes associated with higher reproductive success (i.e., producing more offspring capable of reproducing) might contribute disproportionately to the next generation.

However, these early studies had several limitations. They were based on small sample sizes and, in some cases, involved repeated observations of the same individuals, reducing the robustness of the findings. Moreover, none of the studies investigated whether death-feigning behavior imposes only a temporary physiological cost, or whether there is heritable genetic variation in the trait that would allow it to evolve in response to selection by predators.

In other words, it remained entirely unclear whether death feigning is truly an evolved trait with survival value—i.e., a product of natural selection—rather than simply a nonadaptive physiological reaction to extreme threat.

I contend that the study by Miyatake et al. (2004) was the first to quantitatively demonstrate the adaptive significance of death feigning in relation to survival, supported by robust statistical analyses. (If this claim is incorrect, I welcome direct correspondence for clarification.) As will be elaborated in subsequent chapters, this study established experimental lines selectively bred for long-duration death feigning (L strains) and for minimal or no death-feigning behavior (S strains). When exposed to a natural predator, the jumping spider, individuals from the L strain—those that exhibited death-feigning behavior upon attack—showed a significantly higher short-term survival rate under laboratory conditions compared to individuals from the S strain, which continued to move after predator contact.

Subsequently, Miyatake et al. (2009) conducted an experiment in which one beetle from each of the L and S lineages was placed in a Petri dish with a jumping spider under a triadic condition ($N = 19$). In all trials, it was the S-lineage beetle that was preyed upon first within a fixed observation period. This result suggests that, for predators that target moving prey, death-feigning functions as a "selfish prey"strategy—allowing an individual to enhance its own survival at the expense of a nearby conspecific or heterospecific that continues to move.

Further supporting this notion, Konishi et al. (2020) reported that individuals of *T. castaneum* collected from populations sympatric with a natural predator, the assassin bug, exhibited significantly longer durations of death feigning compared to those from allopatric populations. More recently, Miura et al. (2025) demonstrated that T. castaneum experimentally cohabitated with assassin bugs for five generations displayed significantly longer death feigning durations than conspecifics that had not undergone cohabitation.

Together, these studies provide strong evidence that death feigning serves as an effective antipredator strategy under natural and semi-natural conditions.

In the fire salamander (*Salamandra salamandra*), larval individuals have also been observed to exhibit death-feigning behavior as a means of predator avoidance (Schulte and Caspers 2024). Similarly, studies on tonic immobility in quail have noted that more active individuals are preferentially targeted by predatory cats, indicating that death feigning may serve as an effective last-resort antipredator strategy when direct escape is no longer possible (e.g., Thompson et al. 1981).

A seminal study by Sargeant and Eberhardt (1975) provides early evidence for the adaptive function of death feigning in avian species. Using ducks as the focal species, the study demonstrated that feigning death significantly increased the chances of escape from predatory foxes. Ducks that remained motionless after being captured were reportedly more likely to be released, possibly because the predator relaxed its grip under the assumption that the prey was no longer alive.

The larvae of *Euroleon nostras*, a species of antlion, are sit-and-wait predators that construct pitfall traps in sand to ambush passing prey. Studies on thanatosis in this species have been conducted under the assumption that their primary natural enemies are birds. In a modeling study, Sendova-Franks et al. (2020) and Franks et al. (2021) investigated the effect of death feigning duration on predator redirection. Their results suggested that prolonged durations of immobility confer diminishing returns in terms of diverting predator attention to neighboring antlions within the same patch or to other patches. These studies concluded that the optimal death-feigning strategy is characterized by highly unpredictable durations.

Supporting this, Farji-Brener et al. (2022) reported that the distribution of immobility duration in antlion larvae follows a Poisson distribution with $\lambda = 1$. They argued that such randomness makes the behavior less predictable to predators and thus more effective for avoidance. Once again, this finding underscores that stochastic variability in death-feigning duration enhances its antipredator efficacy.

Thanatosis has also been investigated in the horned beetle *Gnatocerus cornutus*. Matsumura et al. (2020) found that although the frequency of death-feigning behavior decreased with increasing male body and mandible size—suggesting that larger males may rely more on aggression or alternative strategies—the duration of death feigning was not correlated with male size.

Column- I Meet Death Feigning

An earlier version of the following sentences was first published in Japanese in 2022 as "Shinda furi" de ikinobiru by Iwanami Shoten, Publishers, Tokyo. Reproduced by permission of Iwanami Shoten, Publishers.

Until 2000, I worked as a research officer engaged in pest control research in Okinawa. One of the major achievements during this period was the successful eradication of the invasive melon fly (*Bactrocera cucurbitae*)—a serious pest of fruit and vegetable crops—from the Southwestern Islands of Japan through the use of an environmentally friendly technique known as the Sterile Insect Technique (SIT) (Koyama et al. 2004). I was directly involved in this eradication program.

Following the eradication of the melon fly in 1993, attention turned to another invasive pest: the sweet potato weevil (*Cylas formicarius*), a species that targets sweet potatoes. This weevil was first recorded in Okinawa in 1903 and subsequently spread to the Southwestern Islands during the 1950s. The larvae of *C. formicarius* feed on the stems and tubers of sweet potatoes, as well as on morning glories and other members of the Convolvulaceae family.

When attacked, sweet potatoes release a bitter compound known as ipomeamarone. This substance is thought to serve as a defense against fungal pathogens rather than directly against insect herbivory. Due to its extreme bitterness, sweet potatoes damaged by *C. formicarius* are rendered unpalatable to humans and even to livestock, such as pigs.

Until 2000, I worked as a research officer conducting studies on pest management in Okinawa. One of the key achievements during this period was the successful eradication of the invasive melon fly *(Bactrocera cucurbitae*), a major pest of fruits and vegetables, from the Southwestern Islands of Japan through the application of the Sterile Insect Technique (SIT), an environmentally sustainable pest control method (Koyama et al. 2004). I was directly involved in this eradication program.

Following the elimination of the melon fly in 1993, attention shifted to another invasive species, the sweet potato weevil (*Cylas formicarius)*, which attacks sweet potato crops. This species was first recorded in Japan in 1903 in Okinawa and subsequently spread throughout the Southwestern Islands during the 1950s. The larvae of *C. formicarius* feed on the stems and tubers of sweet potatoes, as well as on morning glories and other members of the Convolvulaceae family.

When sweet potatoes are damaged by weevil larvae, they produce a bitter compound known as ipomeamarone. This substance is thought to function as a defense mechanism against fungal infection rather than insect herbivory. Due to its extreme bitterness, infested sweet potatoes are rendered inedible not only to humans but also to livestock such as pigs.

The eradication of the sweet potato weevil (*Cylas formicarius*) has now been confirmed on two southern islands of Japan—Kume Island (Himuro et al. 2022) and Tsuken Island (Ikegawa et al. 2022)—through the use of the Sterile Insect Technique (SIT). The nationwide eradication project targeting this invasive pest began in the early 1990s, with initial efforts focused on elucidating the ecology and distribution of the species. I was involved in both the control efforts and ecological research from the outset of this project. To understand the significance of these achievements, it is important to briefly explain the principles of the SIT. Originally proposed by Knipling (1959), the Sterile Insect Technique is a species-specific pest eradication method that involves mass-rearing the target insect, rendering the individuals sterile—typically through irradiation—and releasing them in large numbers into the wild. When sterile males mate with wild females, the resulting matings are infertile, thereby reducing the reproductive output of the population. Repeated releases of sterile males over multiple generations can drive the target population toward local extinction.

In Japan, the SIT was successfully applied to eradicate the melon fly (*Bactrocera cucurbitae*) from the Southwestern Islands by 1993 (Koyama et al. 2004). Until that time, all successful SIT-based eradications had involved Dipteran species (flies). The eradication of *C. formicarius* on Kume and Tsuken Islands, therefore, represents the first successful application of the SIT to a Coleopteran species (beetles), marking a significant milestone in entomological pest management (Himuro et al. 2022; Ikegawa et al. 2022). During my tenure as a research scientist at the Okinawa Prefectural Agricultural Experiment Station, I collaborated with colleagues to investigate the flight behavior and dispersal capacity of *C. formicarius*, essential parameters for the design of effective SIT programs (Miyatake et al. 1995, 1997, 2000). In addition, I was involved in the development of artificial diets for mass-rearing closely related weevil species, which provided foundational knowledge for the mass production of *C. formicarius* (Shimoji and Miyatake 2002; Urasaki et al. 2024). Successful SIT implementation requires not only the ability to produce and sterilize large numbers of insects, but also a thorough understanding of their dispersal ecology and behavior in the field.

While doing this project research, I was growing sweetpotato weevils, and one day I noticed that when I poked and prodded the insects, they would stop moving in a strange pose. They pretended to be dead. In other words, I saw them feign death. As soon as I saw this insect feign death, I was utterly captivated by the behavior—as if I had fallen in love with it—and that fascination has continued to this day.

2.1.7 Postures of Death Feigning in *Cylas formicarius*

Figure 2.1 presents two images. Can you determine which one shows a weevil exhibiting thanatosis (death feigning), and which one depicts an actually deceased individual? The correct answer is the image on the right. There are three

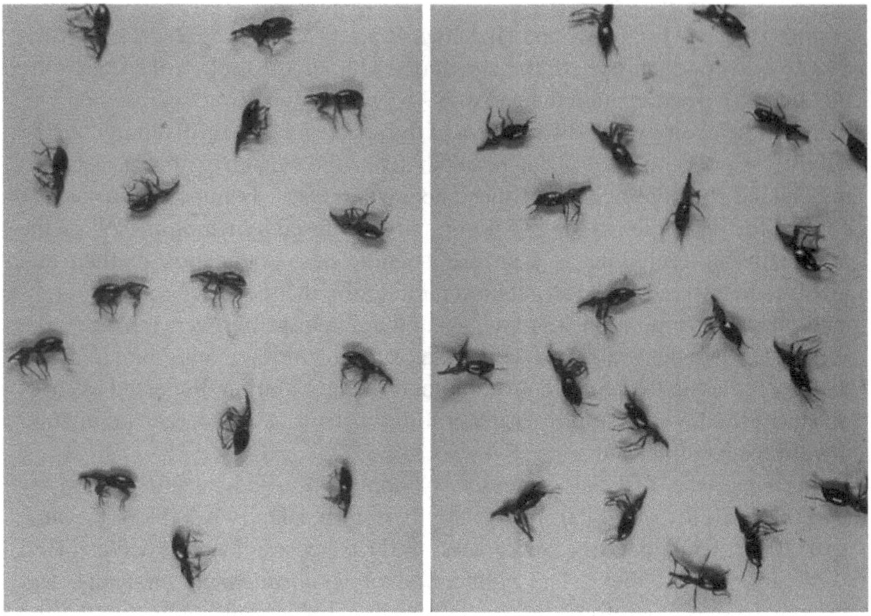

Fig. 2.1 Photographs of an adult sweetpotato weevil (*Cylas formicarius*). The left panel shows a dead adult, whereas the right panel shows an adult exhibiting thanatosis (death-feigning). In the lower portion of the right panel, the weevil is emerging from thanatosis and resuming activity

distinguishing characteristics that can be used to differentiate between a weevil that is truly dead and one that is feigning death, as illustrated in Fig. 2.2.

Point 1: The antennae of a weevil exhibiting death-feigning behavior are symmetrically and deliberately aligned, indicating muscular control.

Point 2: The dorsal surface (back) of a thanatosic weevil often appears unnaturally arched or warped due to tonic muscle tension, whereas the body of a truly dead weevil exhibits a relaxed, natural posture without such distortion.

Point 3: In death-feigning individuals, the hind legs are typically rigid and extended posteriorly in a fixed position. In contrast, all six legs of a dead weevil are flaccid and positioned irregularly due to the absence of muscular tension.

Importantly, these distinctions—observable only in insects, not in deceased humans—suggest that the insect is not in an unconscious state, but rather is actively and deliberately engaging in death-feigning behavior. This realization—that insects may be consciously simulating death—was striking. From that moment, I found myself captivated by the behavioral enigma of thanatosis. I became deeply intrigued by the peculiar stillness of this seemingly absurd posture, and my thoughts were soon consumed by questions surrounding its biological significance.

Under what conditions do insects exhibit death-feigning behavior? What types of stimuli elicit this response, and at what point during an encounter does thanatosis

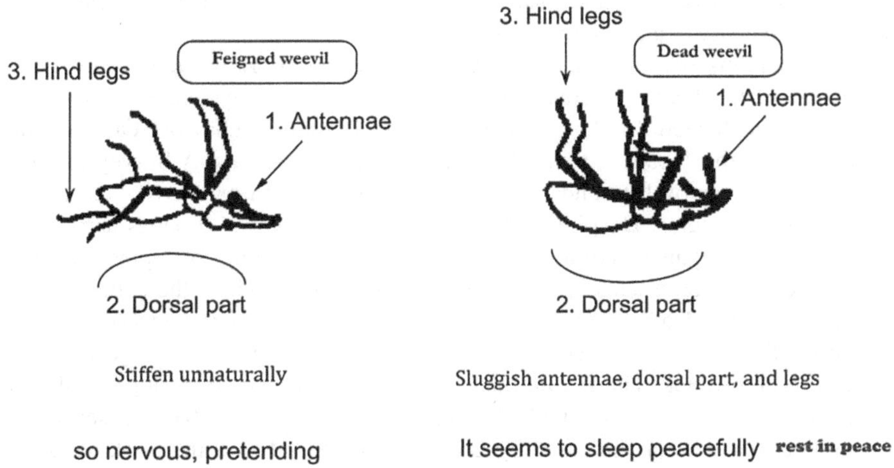

Fig. 2.2 Three diagnostic features for distinguishing between a thanatosis-exhibiting adult (left panel) and a dead adult (right panel) of *Cylas formicarius* are indicated: (1) the position of the antennae, (2) the curvature of the dorsum, and (3) the position of the hind legs

occur? Where and in what ecological contexts is this behavior displayed? Which individuals within a population are more likely to feign death, and to what extent does this tendency vary? What specific stimuli are involved in triggering the response? Why has this behavior evolved—what is its adaptive significance? And finally, how is the behavior performed, and for how long is the immobility maintained? These fundamental questions—the who, what, when, where, why, and how of death feigning—occupied my thoughts as I delved into the mysteries of this fascinating survival strategy.

2.1.8 Is It Science in the First Place?

While I initiated experiments on death-feigning behavior in weevils in 1997, I also conducted a literature review to identify prior research on this phenomenon. If death feigning had already been extensively studied in various insect taxa across different regions of the world, then merely documenting its occurrence in another species would not constitute a novel scientific contribution worthy of publication.

The true value of scientific research lies in uncovering phenomena or insights that no one else in the world has yet explored. The significance of a study is not diminished by the narrowness of its field; if it represents the first discovery of its kind, it constitutes a meaningful contribution to human knowledge. Even when such findings do not yield immediate practical benefits—though societal impact is certainly desirable—they may prove invaluable in the future. A notable example is the serendipitous discovery of the blue-green mold later identified as penicillin, which, despite its accidental origins, ultimately led to a major medical breakthrough and saved countless lives. In this sense, the importance of fundamental research and academic diversity cannot be overstated.

After spending several years visiting bookstores and libraries across Japan during business trips, I came to a realization: as of the 1990s, there were no comprehensive books dedicated to the subject of death-feigning behavior. One notable exception was the Japanese translation of Defence in Animals: A Survey of Anti-Predator Defences by Edmunds (1974), which provides an extensive overview of antipredator strategies. Despite its breadth—spanning over 500 pages across two volumes in the Japanese edition—the topic of death feigning occupies only two pages, one of which is largely composed of illustrations. Thus, the actual textual content devoted to death-feigning behavior amounts to little more than a single page.

During this time, I learned that a comprehensive English-language textbook titled Avoiding Attack: The Evolutionary Ecology of Crypsis, Warning Signals and Mimicry by researchers in the United Kingdom (Ruxton et al. 2004) was scheduled for publication in 2004. Released by Oxford University Press, the book has since become a seminal reference in the study of antipredator strategies, with a second edition published in 2018. It is widely regarded as an authoritative resource that thoroughly explains how various defensive adaptations enhance organismal survival.

Given the scope and prominence of the textbook—and considering that it had been nearly 30 years since the publication of Defence in Animals—I anticipated that feigning death would receive substantial treatment. I assumed that a significant body of research had accumulated in the intervening years, much of which I might have overlooked.

However, my expectations were quickly overturned. Despite the book's 250-page length, feigning death was addressed only briefly—in Chap. 13, Section 6—occupying a mere two pages. Moreover, the authors noted that "death feigning has suffered from recent neglect and a consequent lack of data that meets modern standards of scientific rigour" (Ruxton et al. 2004, p. 199).

Although this realization brought a sense of disappointment, it also led to a clear conviction: whether death feigning contributes to survival remains an open question in evolutionary biology. If no one else was pursuing this line of inquiry, I resolved that I would. At that moment, my scientific curiosity was ignited—the pursuit of knowledge yet unknown to humankind had begun.

2.1.9 The 5W1H of Feigning Death (Thanatosis)

As noted earlier, in 1998, while engaged in a sweet potato weevil (*Cylas formicarius*) eradication project, I developed an interest in the phenomenon of death-feigning behavior in insects. Given that salaries were funded by taxpayers, it was not appropriate to conduct unrelated behavioral observations during official working hours. After considering when such research could be pursued, we decided to carry out preliminary investigations into death feigning during off-duty hours—on weekends or weekday evenings after regular work.

We began by using the sweet potato weevil as a model species to initiate a study on death feigning. The first step involved developing a method for the quantitative analysis of this behavior, aiming to establish a standardized experimental framework for observing and measuring thanatosis in insects.

I promptly visited a local 100-yen shop and purchased several white ceramic coffee cup saucers, along with an inexpensive stopwatch. With these simple materials, the basic tools for the quantitative observation of death-feigning behavior were assembled. The experience was unexpectedly exhilarating—reminiscent of a primary school student's summer holiday science project.

To provide a standardized stimulus to induce death feigning in the weevils, I used a paintbrush that happened to be available. Specifically, I applied the stimulus using the handle end of the brush—the side opposite the bristles. This method proved effective in reliably eliciting death-feigning behavior in adult insects. In retrospect, the brush was unnecessary; a simple stick would have sufficed, as only the handle was used. Nevertheless, this paintbrush came to serve as a reference tool for subsequent experiments.

Through these exploratory trials, a set of operational criteria for assessing death-feigning behavior was established—this formed the basis for conducting what began as a playful experiment in a scientifically rigorous manner. When an insect exhibited tonic immobility (i.e., death feigning) following a single tactile stimulus, a stopwatch was immediately activated to measure the duration until spontaneous resumption of movement. This duration was taken as an indicator of the depth or intensity of the death-feigning response.

In cases where no immobility was observed after the first stimulus, the insect was subjected to up to three successive stimulations. If the individual failed to enter a state of immobility after three stimuli, it was categorized as a "non-responder," and its death-feigning value was set to zero. This quantification was termed the "death-feigning score" (see Miyatake et al. 2004).

The measurement protocol was informed by an earlier study examining geographic variation in death-feigning responses, in which insects were stimulated up to four times (Prohammer and Wade 1981). Notably, this foundational approach has remained largely unchanged and has since become a widely adopted standard among researchers investigating death-feigning behavior in insects worldwide.

Although many students who later inherited and extended this line of research continued to use the same basic methodology, the original paintbrush used for stimulation was eventually replaced by a pair of insect-handling tweezers (model HWC116-10). These tweezers, designed for laboratory use, are soft-tipped and allow for gentle stimulation of delicate beetle specimens without causing harm.

2.1.10 When Does the Beetle Feign Death?

To address the first research question—"Under what conditions do insects feign death in response to external stimulation?"—I conducted a preliminary behavioral assay using adult *Cylas formicarius*. Individual adult weevils were carefully transferred using tweezers from their sweet potato food source onto a white ceramic plate. Each weevil was categorized according to its behavioral state at the moment of transfer: stationary, feeding, or walking. Subsequently, each individual was stimulated with the handle end of a paintbrush, and the death-feigning score was recorded according to the previously established criteria (see Miyatake et al. 2004). The results revealed distinct differences depending on behavioral state. When stimulated, 100% of stationary individuals, regardless of sex, exhibited death-feigning behavior. Similarly, nearly all feeding individuals also entered a state of tonic immobility upon stimulation. In contrast, among walking individuals, the response rate was lower: approximately 30% of males and 50% of females did not feign death, even after three consecutive stimuli. These individuals were assigned a death-feigning score of zero.

These findings suggest that the likelihood of initiating death-feigning behavior varies with the insect's behavioral context, and that individuals in an active locomotor state may have a higher threshold or lower propensity for entering tonic immobility.

When the duration of death-feigning behavior was measured across individuals in different behavioral states, clear differences emerged (Miyatake 2001a). Weevils that were stationary prior to stimulation exhibited the longest durations of tonic immobility, remaining motionless for an average of approximately 8 minutes before resuming movement.

In contrast, individuals that were actively walking at the time of stimulation showed significantly shorter durations, with death-feigning lasting no more than 4 minutes. The feeding individuals displayed intermediate durations, with their mean immobility time falling between those of the stationary and walking groups.

These results indicate that not only the probability of entering a death-feigning state, but also the persistence of that state, is influenced by the weevil's behavioral context at the moment of threat.

Through these experiments, I arrived at an intuitive yet consistent observation: insects appear to exhibit two distinct behavioral modes—an active mode (e.g., walking) and a stationary mode. When an individual is in the active mode, it tends to respond to disturbance by fleeing. In contrast, when in the stationary mode, the insect typically enters a death-feigning state upon stimulation.

This behavioral dichotomy suggests a fundamental principle: the propensity to feign death is strongly state-dependent, and the likelihood of tonic immobility increases when the insect is already immobile at the moment of threat. In this context, death feigning may be understood not as a reaction in place of movement, but rather as a covert behavioral transition—an inverted or alternative form of movement, employed when overt locomotion is not engaged or deemed unwise. This interpretation raises the possibility that thanatosis is not simply a passive absence of movement, but an active behavioral choice within the organism's repertoire—an adaptive, context-specific alternative to escape behavior.

To further investigate the temporal dynamics of death-feigning behavior, I examined when sweet potato weevils (*Cylas formicarius*) are most likely to exhibit tonic immobility. Given that this species is nocturnal, I hypothesized that their responsiveness to threat might vary across the day–night cycle.

A total of 480 adult weevils—comprising 240 males and 240 females—were used in the experiment. The individuals were divided into six groups corresponding to six different time points across a 24-hour period. Starting at 9:00 AM, we tested one group every 4 hours, resulting in testing sessions at 9:00, 13:00, 17:00, 21:00, 1:00, and 5:00. In each session, 30 males and 30 females were examined individually. Each insect was assessed for its death-feigning frequency (whether or not it feigned death upon stimulation) and death-feigning score (duration of immobility, if observed), following the established protocol.

The results revealed a clear sex-based and circadian pattern. Females showed consistently high rates of death-feigning across all time points, including both daytime and nighttime. In contrast, approximately half of the males failed to exhibit death-feigning behavior during the nighttime sessions, particularly at 1:00 and 5:00 AM.

These findings suggest that death-feigning behavior in sweet potato weevils is influenced not only by sex but also by the time of day, potentially reflecting underlying circadian or physiological rhythms that differ between males and females.

When I first presented these findings at a scientific conference, one audience member pointed out a potential confounding factor: the original experiment had been conducted with both male and female weevils present in the same room. The suggestion was that male sweet potato weevils might have altered their behavior due to the presence of female-emitted sex pheromones, particularly during nighttime, when this nocturnal species is most active. It was hypothesized that the males' reduced death-feigning at night might be due to pheromone-driven behavioral arousal overriding their tonic immobility response.

To address this concern, I conducted a follow-up experiment in which only male weevils were present in the laboratory, thereby removing any potential pheromonal influence from females. The death-feigning score was measured at the same six time points used in the original experiment.

The results confirmed the earlier findings: many males still failed to exhibit death-feigning behavior during nighttime hours, even in the absence of females. Furthermore, in those cases where males did feign death at night, the duration of immobility was markedly short—typically lasting only a few minutes before spontaneous recovery.

These findings demonstrate that male sweet potato weevils are intrinsically less likely to feign death at night, independent of female-derived pheromonal cues. This strongly supports the conclusion that circadian or physiological mechanisms underlie the temporal variation in death-feigning behavior in males (Miyatake 2001a).

Why, then, are male sweet potato weevils less likely to feign death at night? Two hypotheses can be proposed to explain this phenomenon. The first is based on a predator–prey interaction strategy. A colleague of mine once dissected a wild rat captured in a sweet potato field and discovered numerous sweet potato weevils in its stomach contents. Since rats are nocturnal predators that rely primarily on olfactory cues to locate prey, feigning death—a visual deception tactic—would likely be ineffective against such predators during nighttime hours. This leads to the hypothesis that death feigning at night may not confer a survival advantage, and therefore, male weevils may have evolved to suppress this behavior during nocturnal periods as an adaptive response to predator sensory ecology. In other words, not feigning death at night may itself be a survival strategy in environments where olfactory-oriented predators dominate.

An alternative hypothesis concerns the reproductive behavior of this species. Specifically, it posits that male sweet potato weevils are so strongly driven to mate during the night that they are unable—or unwilling—to engage in death-feigning behavior.

Female sweet potato weevils are known to release sex pheromones nocturnally to attract potential mates. Males, in turn, possess prominently developed antennae, which are considerably larger than those of females, presumably to enhance their sensitivity to these chemical cues. In such a critical mating window, it is plausible that remaining motionless in a death-feigning state could reduce a male's chances of locating and mating with a receptive female. From an evolutionary perspective, a male that prioritizes reproductive opportunity over antipredator defense may ultimately leave more offspring.

Interestingly, females continue to feign death effectively at night, suggesting a sex-specific trade-off between reproduction and survival (see Miyatake 2001a). Although still speculative, I now consider that sexual selection and reproductive pressure may be the primary drivers underlying the suppression of death-feigning behavior in males during nocturnal periods.

In summary, these findings suggest that in this nocturnal species, death-feigning behavior is suppressed during the active nighttime period, whereas it is more readily expressed during the daytime rest phase. This pattern indicates that thanatosis serves as a context-dependent antipredator strategy, employed primarily when the insect is inactive and not engaged in reproductive or foraging activities. Based on these results, I submitted a manuscript to the International Journal of Insect Behaviour, which was published as my first scientific paper on death feigning (Miyatake 2001a). More recently, new evidence has emerged indicating that male sweet potato weevils in a state of tonic immobility can be aroused from death feigning by exposure to female sex pheromones. Specifically, males respond to olfactory cues from sexually mature females, resuming activity even when they had previously entered a death-feigning state (Himuro and Miyatake 2024). This finding further supports the hypothesis that reproductive motivation can override antipredator behavior, revealing a dynamic trade-off between survival and mating effort.

After I transferred from Okinawa to Okayama University, a junior colleague who had joined the sweet potato weevil research team initiated a new line of investigation examining the relationship between mating behavior and death feigning. This work led to two published studies (Kuriwada et al. 2009, 2011).

In the first study, it was shown that mating experience influenced the duration of death-feigning behavior in sweet potato weevils (Kuriwada et al. 2009). Specifically, mating did not affect death-feigning duration in males, but in females, postmating fertilization significantly reduced the duration of tonic immobility. Interestingly, females that had mated but were not fertilized tended to exhibit longer death-feigning durations.

These results suggest that mating experience—and more precisely, reproductive physiological status—modulates death-feigning behavior, at least in females. Taken together, the findings indicate that reproductive state and sex-specific strategies are intricately linked to the expression of thanatosis in this species. It is becoming increasingly clear that the ecological and evolutionary dynamics of feigning death cannot be fully understood without considering sex differences in reproductive investment and motivation.

2.1.11 Energetic State Constrains Antipredator Behavior: Death Feigning Is Suppressed Under Nutritional Stress

A question I have frequently been asked is: "Do organisms ever die while feigning death?"

At first glance, this seems paradoxical—if tonic immobility is a controlled, active antipredator behavior, it is difficult to imagine an organism dying while deliberately maintaining such a state. However, the question raises a valid point, particularly when considering extreme physiological conditions such as starvation. This led me to examine the relationship between nutritional status and death-feigning behavior in more detail.

To explore this, I conducted experiments using the sweet potato weevil (*Cylas formicarius*), a model species in my research on thanatosis. The study aimed to determine whether prolonged food deprivation would alter an insect's ability to feign death—or, in extreme cases, whether the boundary between tonic immobility and actual death might blur under metabolic stress.

This investigation not only addressed the question of whether death feigning could be sustained under energy-limited conditions, but also provided insights into the energetic cost and physiological requirements underlying this seemingly passive behavior.

Although somewhat harsh from the perspective of the insects, this experiment involved withholding food (sweet potatoes) from sweet potato weevils to assess the effects of starvation on their death-feigning behavior. Before initiating the experiment, I had several concerns—most notably, the well-established observation across taxa that males and females often differ in starvation tolerance. In many species, females are known to accumulate greater energy reserves, such as body fat, to support future reproductive investment, such as egg production.

To address this, I first examined sex-specific differences in starvation tolerance in *Cylas formicarius*. The results aligned with prior expectations based on physiological differences between the sexes:

Male weevils exhibited lower starvation resistance: the majority died within 10 days of food deprivation, and complete mortality was observed by day 15. Female weevils showed significantly greater tolerance: almost all individuals survived up to 15 days without food, and 100% mortality was not reached until day 23.

These findings confirm a pronounced sexual dimorphism in starvation resistance, likely attributable to greater energy reserves in females, which are thought to support future reproductive investment.

These findings (Miyatake 2001b) also demonstrate that females possess significantly greater resistance to starvation than males, likely due to higher endogenous energy reserves associated with reproductive physiology. This sex-specific physiological asymmetry should be carefully considered when interpreting behavioral modifications—such as variations in the expression or duration of death-feigning behavior—under conditions of nutritional stress.

Given the previously observed sex differences in starvation tolerance, I conducted a follow-up experiment to examine how nutritional state affects the expression of death-feigning behavior in male and female sweet potato weevils (*Cylas formicarius*). A total of 150 males and 150 females were used. For each sex, 50 individuals were placed in a container with sweet potatoes, and the experiment was replicated three times under identical conditions.

In each trial, food was removed simultaneously from all containers, and the frequency and intensity of death-feigning behavior were assessed at various time points by gently stimulating individuals during the daytime, when death feigning is normally most pronounced in this species.

The results were consistent across all three replicates: By Day 3 of food deprivation, approximately 60% of males no longer exhibited death-feigning behavior. In contrast, 100% of females still feigned death after the same period. After 6 days of starvation, nearly half of the males failed to feign death, whereas the vast majority of females continued to do so. On Day 15, all males had died, while only 25% of females were still capable of death feigning. Moreover, among males that did feign death after 6 days of starvation, the duration of immobility was markedly reduced, and many resumed activity quickly in search of food, resulting in low death-feigning scores. These findings indicate that nutritional stress strongly suppresses death-feigning behavior in males, and to a lesser extent in females, reflecting both their physiological differences in energy reserves and potentially sex-specific prioritization of survival strategies under resource-limited conditions.

The experiments have shown that weevils cannot feign death when they are hungry. After all, feigning death was intentional, and no insect would play dead for long when it was about to face death. Starving insects have no choice but to become active and cannot go into a feign-dead mode. The results were promptly submitted to the Journal of the Entomological Society of America and published in 2001 (Miyatake 2001b). The phenomenon that starving a beetle reduces the time it feigns death has also been replicated by another research groups using other weevil species, which supports this phenomenon (Li et al. 2018).

2.1.12 Beetles Can Only Feign Death if They Can Afford to Do So

In the autumn of 2000, as I was preparing to finalize and submit the results of my study on starvation and death-feigning behavior in the sweet potato weevil (*Cylas formicarius*) to the Journal of the Entomological Society of America, I was appointed to an open faculty position at Okayama University.

The sweet potato weevil is an invasive pest species that originated in Southeast Asia and has spread across the southwestern islands of Japan, particularly Okinawa. However, it is not present on the main islands of Honshu, including Okayama. According to Japanese law, transporting this regulated pest species from Okinawa to other regions is strictly prohibited, even for research purposes. Although a special containment facility can be used under permit, I decided against maintaining live populations of *C. formicarius* at Okayama University due to the risks associated with unintentional escape—particularly in a university setting with frequent student traffic.

A potential outbreak of this species in Honshu would be a serious agricultural concern, especially given the proximity to major sweet potato cultivation areas such as those producing the renowned "Naruto Kintoki" variety. In light of these risks, I made the difficult but necessary decision to discontinue experimental work on death-feigning behavior in the sweet potato weevil following my relocation.

Following the discontinuation of experimental work on the sweet potato weevil, I turned to another beetle species for continued research on death-feigning behavior: the adzuki bean beetle (*Callosobruchus chinensis*). This species, which I had been keeping in Okinawa as a research organism—and, informally, as something of a "pet"—was legally and safely transported to Okayama (Fig. 2.3).

As its name implies, the adzuki bean beetle is a common pest of legumes, including adzuki beans and cowpeas. The adult beetles oviposit on dry beans in the autumn, and the emerging larvae bore into the beans and develop internally, feeding as they grow.

Interestingly, this beetle also exhibits death-feigning behavior, though its posture differs from that of the sweet potato weevil. While *C. formicarius* typically flips onto its back and tucks in its legs, the adzuki bean beetle shows a more compact form of immobility. Upon "revival" from the tonic immobility state, the beetle

Fig. 2.3 Death-feigning posture of the aduzki bean beetle, *Callosobruchus chinensis* (Hozumi and Miyatake 2005)

cautiously moves only its long antennae at first, apparently to assess its surroundings before standing up—a behavior that, from a human perspective, appears unexpectedly charming.

Around the same time, I began supervising undergraduate students for their graduation research projects at Okayama University. When one student approached me for advice on a thesis topic, I proposed a study on death-feigning behavior in the adzuki bean beetle (*Callosobruchus chinensis*).

I still vividly recall the student's reaction—a blend of surprise, confusion, and mild disappointment, as if to say, "Feigning death? That's going to be my graduation thesis?" For a student who likely entered the Faculty of Agriculture with aspirations of working on applied topics such as pest management or crop production, a behavioral ecology project on tonic immobility may have seemed an unexpected and perhaps even puzzling choice.

Nevertheless, this encounter marked the beginning of what would become a long and fruitful line of undergraduate-led investigations into the adaptive significance and mechanisms of death feigning—a topic that, while unconventional at first glance, is rich in evolutionary and ecological insight.

Given that many students likely entered the Faculty of Agriculture with the expectation of engaging in applied research—such as pest control, crop improvement, or agricultural biotechnology—it is perhaps unsurprising that a thesis project focused on thanatosis in a small bean-infesting beetle may have seemed, at first glance, rather unexpected. Yet such initial reactions often reflect a narrower view of agricultural science, and the project served as an opportunity to broaden perspectives: even a seemingly obscure behavior like death feigning offers deep insights into evolutionary adaptation, predator–prey interactions, and the ecological underpinnings of pest species.

However, once I explained to the students how fascinating the study of death-feigning behavior in beetles could be, and we began outlining the structure of their thesis projects, their initial skepticism gave way to genuine interest.

To begin experimentation, we visited a local 100-yen shop in Okayama and procured the basic tools: a white ceramic plate, a stopwatch, and a paintbrush—the same simple apparatus used in earlier studies. The first project undertaken by the undergraduate student focused on whether there are sex-based differences in death-feigning behavior in the adzuki bean beetle (*Callosobruchus chinensis*).

This beetle species offers numerous advantages as an experimental model. When multiple eggs are laid on a single adzuki bean, larvae compete for limited resources within the seed, resulting in a smaller body size compared to larvae raised in isolation. This natural intraspecific competition has made the species a popular subject in studies of resource-based developmental plasticity.

Moreover, experimental manipulation is straightforward: the number of larvae developing per bean can be easily controlled by selectively removing excess eggs using a scalpel or fine knife. With the availability of a precision electronic balance at the university, even subtle differences in adult body mass could be quantified accurately—an important consideration, as body size is known to influence both defensive behaviors and reproductive strategies in many insect species.

Once adzuki bean beetles (*Callosobruchus chinensis*) emerge as adults, their growth ceases, as their bodies are fully sclerotized and enclosed within a rigid exoskeleton. Consequently, adult body size and mass are fixed at the time of emergence, making these traits ideal for investigating correlations with behavioral characteristics such as death feigning. To explore the relationship between body size and death-feigning behavior, we weighed newly emerged adults using a precision electronic balance. Two experimental groups were established based on larval crowding conditions: One group developed from beans on which more than 10 eggs had been laid, resulting in intense larval competition and smaller adult body size. The other group developed from beans with a single egg, ensuring exclusive access to resources and a larger adult body size. This setup enabled us to examine whether variations in developmental conditions and resulting body size influenced the frequency and duration of death-feigning behavior.

The results revealed a clear positive correlation between adult body weight and the duration of death-feigning behavior: larger individuals remained immobile for significantly longer periods than smaller ones. To our knowledge, this was the first empirical evidence globally demonstrating an association between body mass and thanatosis in insects. To confirm the reliability of this finding, the experiment was replicated under identical conditions. The results were consistent, reinforcing the validity of the observed trend.

Interestingly, we also found that smaller-bodied adults exhibited only about half the lifespan of their larger counterparts. One possible explanation is that individuals with limited energy reserves may prioritize immediate survival strategies, such as fleeing, rather than engaging in prolonged immobility. In other words, they may not have the metabolic luxury to "play dead." While this remains a working hypothesis, it highlights the need for further investigation into the physiological and ecological constraints underlying antipredator behaviors. The findings from this undergraduate research were published in the Journal of Insect Behavior in 2005 (Hozumi and Miyatake 2005), marking the first peer-reviewed report linking body size to death-feigning behavior in this species.

2.1.12.1 Can't Do It in the Heat

Although it is less commonly seen in Japan today, it was once not unusual to find flies and other insects immobilized inside supermarket freezer sections. This everyday observation reflects a well-known biological phenomenon: insects, as ectothermic and metamorphic organisms, become immobile at low temperatures.

Based on previous observations, we hypothesized that under low-temperature conditions—when insects are unable to move—they may be involuntarily forced into a state resembling death feigning. Conversely, as ambient temperature increases and metabolic activity rises, insects may become more active and less likely to exhibit death-feigning behavior. This idea, while speculative, seemed plausible in light of known ectothermic physiology.

I believe it was around 6 years after I joined Okayama University when we had an opportunity to test this hypothesis. At that time, we acquired a large, walk-in environmental chamber capable of finely controlling temperature across a wide range. This allowed us not only to manipulate breeding and experimental conditions with precision but also to observe insect behavior directly under various thermal environments.

Although the hypothesis appeared intuitive, we adhered to the principle that even seemingly obvious phenomena require empirical validation through controlled experimentation.

To test this hypothesis, we placed adult adzuki bean beetles (*Callosobruchus chinensis*) in walk-in environmental chambers set to five different constant temperatures: 15 °C, 20 °C, 25 °C, 30 °C, and 35 °C (Miyatake et al. 2008). Although I had assistance from a graduate student in handling the insects, I personally conducted all of the behavioral observations to ensure consistency and accuracy. As predicted, the proportion of adults exhibiting death-feigning behavior decreased with increasing temperature. At 15 °C and 20 °C, nearly 100% of the beetles feigned death upon stimulation. In contrast, at 30 °C and 35 °C, only about 50% of individuals did so. Moreover, among those that did feign death, the duration of immobility was significantly reduced at higher temperatures, resulting in lower death-feigning scores.

These findings support the hypothesis that lower temperatures, which suppress locomotor activity, facilitate the expression of tonic immobility, whereas higher temperatures, which increase activity levels, inhibit it. In short, when beetles are physiologically active due to warmer temperatures, they are less likely—and perhaps less able—to feign death.

To confirm the reproducibility of our findings, we conducted a similar experiment using a closely related species, *Callosobruchus maculatus*, which was kindly provided by a colleague. As with C. chinensis, we found that death-feigning behavior was strongly influenced by ambient temperature: at higher temperatures, individuals were less likely to feign death, and those that did exhibited shorter durations of immobility. In contrast, at lower temperatures, the majority of beetles displayed prolonged death feigning. These consistent results across species reinforce the conclusion that temperature modulates the expression of tonic immobility in bruchid beetles. The full findings were published in Physiological Entomology, a journal of the Royal Entomological Society, in 2008 (Miyatake et al. 2008).

At the risk of sounding somewhat boastful, I could not help but recall a passage from The Insect Chronicles by Jean-Henri Fabre (1910), in which he lamented:

> The duration of the immobile posture varies terribly, even when done on the same day, in the same atmospheric conditions, and with the same experimental materials. I have never been able to find any cause for shortening or lengthening it.

Reading these words, I found myself wishing that Dr. Fabre could witness the results of our systematic experiments. Perhaps, he would be pleased to see that, more than a century later, modern experimental tools and controlled conditions have allowed us to identify temperature—and even body size—as key factors modulating death-feigning behavior in insects.

2.1.13 Active and Stationary Modes

To summarize the findings so far, beetles are less likely to exhibit death-feigning behavior when they are in an active state—that is, when they are walking, feeding, mating, hungry, exposed to high temperatures, or are small in body size. In contrast, death feigning is more frequently observed when beetles are resting, satiated, exposed to cooler temperatures, or have grown to a larger body size.

These results support the idea that insects alternate between two behavioral modes—active and stationary. When in the active mode, insects tend to flee in response to a threat, while those in the stationary mode are more likely to enter tonic immobility as a defensive strategy. The transition between these modes is governed by internal physiological states and environmental conditions, including temperature, hunger, and developmental history.

But an important question remains: Does feigning death actually enhance survival and contribute to evolutionary fitness?

This is a question that neither Darwin nor Fabre could answer conclusively. In Chap. 5, we will explore an experiment designed to directly address whether death feigning can evolve as an adaptive trait. Please stay tuned.

2.1.14 Previous Review Papers

A number of essays and review articles have explored death-feigning behavior, also referred to as tonic immobility or thanatosis, across various taxa and disciplinary perspectives. One of the earliest technical works to address death feigning as a predator avoidance strategy is Defence in Animals by Edmunds (1974). Later, Ruxton et al. (2004) introduced thanatosis in the context of antipredator adaptations in their influential book Avoiding Attack: The Evolutionary Ecology of Crypsis, Warning Signals, and Mimicry. In the second edition of this work, a dedicated chapter on thanatosis was added (Ruxton et al. 2018), reflecting the growing academic interest in the topic. Humphreys and Ruxton (2018) have explained thanatosis (death feigning) from a variety views, including physiological mechanisms, adaptive significance, and cost–benefit relationships, and introduced the example with broad sense in ecology, physiology, and coevolutionary aspects (Also see Rogers and Simpson 2014).

From a psychological and ethological standpoint, Reese et al. (1984) reviewed tonic immobility in birds and other animals, while Skelhorn (2018) provided an essay addressing thanatosis through the lens of predator–prey dynamics. Monsó (2024) has published a book titled "Playing Possum: How Animals Understand Death" for a broad audience. This book illustrates that the concept of "death" is not unique to humans but is widely present across the animal kingdom. It explores how animals, ranging from insects to mammals, perceive and respond to death. Incorporating recent advances in behavioral science and comparative psychology, this collection of essays also addresses death-feigning behavior, offering insights into a diverse array of death-related behaviors observed in animals.

More recently, a comprehensive volume of papers on tonic immobility—ranging from neurobiological mechanisms to adaptive significance—was published by Carli and Farabollini (2022). In insects, mechanisms and functions of death feigning behaviours have been published as special paper's collection in a book (Sakai 2021a, b) including Pflüger et al. (2021), Honma (2021), Nakano (2021), Takanashi and Kojima (2021), Nishino and Sakai (2021), Büschges et al. (2021), Namiki (2021), Yoshida (2021), and Miyatake (2021). The most up-to-date review of tonic immobility across animal species appears in Woodruff (2025), which synthesizes current knowledge and identifies key questions for future research.

References

Acheampong A, Mitchell BK (1997) Quiescence in the colorado potato beetle, *Leptinotarsa decemlineata*. Entomol Exp Appl 82:83–89

Arduino PJ Jr, Gould AJ (1984) Is tonic immobility adaptive? Anim Behav 32:921–923

Asakura M, Matsumura K, Ishihara R, Miyatake T (2022) Freezing or death feigning? Beetles selected for long death feigning showed different tactics against different predators. Ecol Evol 12:e8533

Büschges A, Pflüger H, Bässler U (2021) Catalepsy and twig mimesis in insects and its neural control. In: Sakai M (ed) Death-feigning in insects, entomology monographs. Springer Nature Singapore Pte Ltd, Singapore, pp 135–143

Carli G, Farabollini F (2022) Defence from invertebrates to mammals: focus on tonic immobility, Progress in Brain Research, vol 271. Elsevier, Amsterdam, p 337

Complete Fabre Insect Chronicle 7, 1993 Iwanami Shoten

Darwin CR (1883) A posthumous essay on instinct. In: Romanes GJ (ed) Mental evolution in animals. With a posthumous essay on instinct by Charles Darwin. Kegan Paul Trench & Co., London, pp 188–189, 196, 198–199, 301–302, 355–384

Edmunds M (1974) Defence in animals: a survey of anti-predator defences. Longman, London

Fabre JH (1910) Souvenirs Entomologiques. Delgrave Paris, Paris

Farji-Brener AG, Cajas A, Campos-Alvarado A, Castaño-Diaz M, Estévez-Haro M, Piñanez-Espejo YM (2022) Understanding unpredictability: factors influencing how long antlion larvae play dead as an antipredator behaviour. Anim Behav 194:199–204

Franks NR, Worley A, Sendova-Franks AB (2021) Hide-and-seek strategies and post-contact immobility. Biol Lett 17:20200892

Gilman TT, Marcuse FL, Moore AU (1950) Animal hypnosis: a study in the induction of tonic immobility in chickens. J Comp Physiol Psychol 43:99–111

Himuro C, Miyatake T (2024) Mate search or predation avoidance? Sex pheromone interrupts death feigning of males in the sweet potato weevil *Cylas formicarius*. J Ethol 42:169–176

Himuro C, Kohama T, Matsuyama T, Sadoyama Y, Kawamura F, Honma A, Ikegawa Y, Haraguchi D (2022) First case of successful eradication of the sweet potato weevil, Cylas formicarius (Fabricius), using the sterile insect technique. PLoS One. https://doi.org/10.1371/journal.pone.0267728

Holmes JS (1906) Death-feigning in *Ranatra*. J Comp Neurol Psychol 16:200–216

Honma A (2021) The function of tonic immobility: review and prospectus. In: Sakai M (ed) Deathfeigning in insects, entomology monographs. Springer Nature Singapore Pte Ltd, Singapore, pp 27–37

Honma A, Oku S, Nishida T (2006) Adaptive significance of death feigning posture as a specialized inducible defence against gape-limited predators. Proc Royal Soc B 273:1631–1636

Hozumi N, Miyatake T (2005) Body-size dependent difference in death-feigning behavior of adult *Callosobruchus chinensis*. J Insect Behav 18:557–566

Humphreys RK, Ruxton GD (2018) A review of thanatosis (death feigning) as an anti-predator behaviour. Behav Ecol Sociobiol 72:22

Ikegawa Y, Kawmura F, Sadoyama Y, Kinjo K, Haraguchi D, Honma A, Himuro C, Matsuyama T (2022) Eradication of sweetpotato weevil, *Cylas formicarius*, from Tsuken Island, Okinawa, Japan, under transient invasion of males. J Appl Entomol 146:850–859

Katsuki M, Kiyose K, Okada K (2024) Predation facilitates evolution of sex-specific antipredator defences in a sexually dimorphic beetle. Anim Behav 215:81–88

Keyakizaka 46 Official You Tube Site (2025) "Shindafuri" Music Video. https://www.youtube.com/watch?v=WNH_ReRMt5k

Klemm WR (1977) Identity of sensory and motor systems that are critical to the immobility reflex ("animal hypnosis"). Psychol Rec 1:145–159

Knipling EF (1959) Sterile-male method of population control: successful with some insects, the method may also be effective when applied to other noxious animals. Science 130:902–904

Konishi K, Matsumura K, Sakuno W, Miyatake T (2020) Death feigning as an adaptive anti-predator behaviour: further evidence for its evolution from artificial selection and natural populations. J Evol Biol 33:1120–1128

Koyama J, Kakinohana H, Miyatake T (2004) Eradication of the melon fly, *Bactrocera cucurbitae*, in Japan: importance of behavior, ecology, genetics, and evolution. Annu Rev Entomol 49:331–349

Kuriwada T, Kumano N, Shiromoto K, Haraguchi D (2009) Copulation reduces the duration of death-feigning behaviour in the sweetpotato weevil, *Cylas formicarius*. Anim Behav 78:1145–1151

Kuriwada T, Kumano N, Shiromoto K, Haraguchi D (2011) Age-dependent investment in death-feigning behaviour in the sweetpotato weevil *Cylas formicarius*. Physiol Entomol 36:149–154

Li H, Zhang G, Ji Y, Wen J (2018) Effects of starvation on death-feigning in adult *Eucryptorrhynchus brandti* (Coleoptera: Curculionidae). Ethology 125:645–651

Matsumura K, Yumise K, Fujii Y, Hayashi T, Miyatake T (2020) Anti-predator behaviour depends on male weapon size. Biol Lett 16:20200601

Miura Y, Takii H, Matsumura K, Miyatake T (2025) Experimental evolution of prey by predation pressure: death-feigning duration becomes longer in the presence of predators. Appl Entomol Zool 60:257. https://doi.org/10.1007/s13355-025-00913-z

Miyatake T (2001a) Diurnal periodicity of death-feigning in *Cylas formicarius* (Coleoptera: Brentidae). J Insect Behav 14:421–432

Miyatake T (2001b) Effects of starvation on death-feigning in adults of *Cylas formicarius* (Coleoptera: Brenridae). Ann Entomol Soc Am 94:612–616

Miyatake T (2021) Environmental, physiological, and genetic effects on tonic immobility in beetles. In: Sakai M (ed) Death-feigning in insects, entomology monographs. Springer Nature Singapore Pte Ltd, Singapore, pp 39–53

Miyatake T, Kawasaki K, Kohama T, Moriya S, Shimoji Y (1995) Dispersal of male sweetpotato weevils (Coleoptera: Curculionidae) in fields with or without sweet potato plants. Environ Entomol 24:1167–1174

Miyatake T, Moriya S, Kohama T, Shimoji Y (1997) Dispersal potential of male sweetpotato weevils (Coleoptera: Brentidae) over land and water. Environ Entomol 26:272–276

Miyatake T, Kohama T, Shimoji Y, Kawasaki K, Moriya S, Kishita M, Yamamura K (2000) Dispersal of released male sweetpotato weevil, *Cylas formicarius* (Coleoptera: Brentidae) in different seasons. Appl Entomol Zool 35:441–449

Miyatake T, Katayama K, Takeda Y, Nakashima A, Sugita A, Mizumoto M (2004) Is death-feigning adaptive? Heritable variation in fitness difference of death-feigning behaviour. Proc R Soc Lond Ser B 271:2293–2296

Miyatake T, Okada K, Harano T (2008) Negative relationship between ambient temperature and death-feigning intensity in adult *Callosobruchus maculatus* and *Callosobruchus chinensis*. Physiol Entomol 33:83–88

Miyatake T, Nakayama S, Nishi Y, Nakajima S (2009) Tonically immobilized selfish prey can survive by sacrificing others. Proc R Soc B 276:2762–2767

Monsó S (2024) Playing possum: how animals understand death. Princeton University Press

Nakano R (2021) Ultrasound-induced freezing response in moths. In: Sakai M (ed) Death-feigning in insects, entomology monographs. Springer Nature Singapore Pte Ltd, Singapore, pp 55–72

Nakayama S, Miyatake T (2009) Genetic trade-off between abilities to avoid attack and to mate: a cost of tonic immobility. Biol Lett 6:18–20

Namiki S (2021) Descending neuron for freezing behavior in *Drosophila melanogaster*. In: Sakai M (ed) Death-feigning in insects, entomology monographs. Springer Nature Singapore Pte Ltd, Singapore, pp 145–157

Nishino H (2004) Motor output characterizing thanatosis in the cricket *Gryllus bimaculatus*. J Exp Biol 207:3899–3915

Nishino H, Sakai M (1996) Behaviorally signifcant immobility so called of thanatosis in the cricket *Gryllus bimaculatus* DeGeer. J Comp Physiol 179:613–624

Nishino H, Sakai M (2021) Tonic immobility in a cricket: behavioral characteristics, neural substrate, and functional significance. In: Sakai M (ed) Death-feigning in insects, entomology monographs. Springer Nature Singapore Pte Ltd, Singapore, pp 93–108

Nishino H, Sakai M, Field LH (1999) Two antagonistic functions of neural groups of the femoral chordotonal organ underlie thanatosis in the cricket *Gryllus bimaculatus* DeGeer. J Comp Physiol A Neuroethol Sens Neural Behav Physiol 185:143–155

Ohno T, Miyatake T (2007) Drop or fly? Negative genetic correlation between death-feigning intensity and flying ability as alternative anti-predator strategies. Proc R Soc B 274:555–560

Pflüger H, Büschges A, Bässler U (2021) Catalepsy and twig mimesis in insects and its neural control. In: Sakai M (ed) Death-Feigning in Insects, Entomology Monographs. Springer Nature Singapore Pte Ltd, Singapore, pp 15–21

Prohammer LA, Wade MJ (1981) Geographic and genetic variation in death-feigning behavior in the flour beetle, *Tribolium castaneum*. Behav Genet 11:395–401

Reese WG, Angel C, Joseph EO, Newton MD (1984) Immobility reactions: a modified classification. Pavlov J Biol Sci 19:137–142

Robertson TB (1904) On the 'Sham-death' reflex in spiders. J Physiol Lond 31:410–417

Rogers SM, Simpson SJ (2014) Thanatosis. Curr Biol 24:R1031–R1033

Ruxton G (2006) Grasshoppers don't play possum. Nature 440:880. https://doi.org/10.1038/440880a

Ruxton GD, Sherratt TN, Speed MP (2004) Avoiding attack. Oxford University Press

Ruxton GD, Allen WL, Sherratt TN, Speed MP (2018) Avoiding attack: the evolutionary ecology of crypsis, aposematism, and mimicry. Oxford University Press, Oxford

Sakai M (2021a) Death-feigning in insects: mechanism and function of tonic immobility. Springer, Tokyo, p 189

Sakai M (2021b) Freezing and tonic immobility: their definitions and naming. In: Sakai M (ed) Death-feigning in insects, entomology monographs. Springer Nature Singapore Pte Ltd, pp 1–14

Sargeant AB, Eberhardt LE (1975) Death feigning by ducks in response to predation by red foxes (*Vulpes fulva*). Am Midl Nat 94:108–119

Schulte L, Caspers BA (2024) Tonic immobility behaviour does not differ between fire salamander larvae from ponds and streams. Ecol Evol 14:e11211

Sendova-Franks AB, Worley A, Franks NR (2020) Post-contact immobility and half-lives that save lives. Proc R Soc Lond B Biol Sci 287:20200881. https://doi.org/10.1098/rspb.2020.0881

Shimoji Y, Miyatake T (2002) Adaptation to artificial rearing during successive generations in the West Indian sweetpotato weevil, *Euscepes postfasciatus* (Coleoptera: Curuculionidae). Ann Entomol Soc Am 95:735–739

Skelhorn J (2018) Avoiding death by feigning death. Curr Biol 28:R1121–R1142

Takanashi T, Kojima W (2021) Vibration-induced immobility in coleopteran insects. In: Sakai M (ed) Death-feigning in insects, entomology monographs. Springer Nature Singapore Pte Ltd, Singapore, pp 73–92

Thompson RKR, Foltin RW, Sweet BA, Graves CA, Lowitz CE (1981) Tonic immobility in Japanese quail can reduce the probability of sustained attack by cats. Anim Lean Behav 9:145–149

Urasaki K, Matsumura K, Miyatake T (2024) Spatio-temporal distribution of adults and eggs of the West Indian sweetpotato weevil *Euscepes postfasciatus* (Coleoptera: Curculionidae) on sweet potato stems. Appl Entomol Zool 59:117–126

Woodruff ML (2025) Tonic immobility and phenomenal consciousness in animals: a review. Front Psychol. https://doi.org/10:3389/fpsyg.2025.1509999

Yoshida M (2021) Immobility behaviors in fish: a comparison with other vertebrates. In: Sakai M (ed) Death-feigning in insects, entomology monographs. Springer Nature Singapore Pte Ltd, Singapore, pp 159–178

Yoshikawa T (1989) Sangoku-shi. Kodansha, Tokyo, p 512. (in Japanese)

Overview of Each Taxon

3

Abstract

Death-feigning behavior has been reported in a wide range of animal taxa. Taxa of animals in which death feigning, tonic immobility, or related immobility behaviors have been observed or reported are listed in Table 3.2. This chapter provides an overview of the types of immobility behaviors documented and the research conducted for each taxonomic group. The discussion begins with arthropods, followed by vertebrates. Death-feigning behavior has been reported in three animal phyla—Mollusca, Arthropoda, and Chordata—and has been observed in animals belonging to 12 classes and 49 orders. In this chapter, for each order, I provide the species in which death-feigning behavior has been reported, the corresponding references, the terminology used to describe the behavior, and the life stage at which the behavior was observed. In other words, if readers happen to observe death-feigning behavior in a taxonomic group not included in this list, such an observation would be of scientific value and worth reporting in an academic journal.

Keywords

Mollusca · Arthropoda · Chordata · Taxonomic group · Stage · Vertebrate · Invertebrates

Table 3.1 presents a list of taxonomic groups in which death-feigning behavior has been observed in the animal kingdom. For more detailed descriptions of death-feigning observed in each taxonomic group, please refer to the following text.

© The Author(s), under exclusive license to Springer Nature Singapore Pte Ltd. 2026

T. Miyatake, *Death Feigning*, Ecological Research Monographs,
https://doi.org/10.1007/978-981-95-5319-8_3

Table 3.1 Orders in which tonic immobility was observed among animals (Animalia)

Phylum	Subphylum	Class	Subclass	Order	Statge	References
Mollusca		Gastropoda	Caenogastropoda	Neogastropoda	Adults	Ulate Gómez et al. (2024)
Arthropoda	Chelicerata	Arachnida	Scorpiomoroha	Scorpiones	Adults	Lira et al. (2020) and Trian et al. (2022)
			Araneomorpha	Opiliones	Adults	Segovia et al. (2019a)
		Crustacea	Branchiopoda	Diplostraca	Adults	Yamada and Urabe (2021)
			Malacostraca	Isopoda	Adults	Quadros et al. (2012), Suzuki and Futami (2017), and Igarashi and Wada (2024)
				Decapoda	Adults	Pereyra et al. (1999), Coutinho et al. (2013), and Scarton et al. (2009)
				Araneae	Adults	Bilde et al. (2006), Hansen et al. (2008), and Liu et al. (2024)
				Acarina	Adults, nymphs	Oyen et al. (2021)
		Diplopoda	Chilognatha	Polydesmida	Adults	Bouzan et al. (2023)
		Chilopoda	Pleurostigmomorpha	Scolopendromorpha	Adults	Vega-Román et al. (2024)
		Insecta		Odonata	Larvae	Gyssels and Stoks (2005)

					Asults	Khelifa (2017) and Wildermuth et al. (2019)
				Plecoptera	Larvae	Moore and Williams (1990)
				Orthoptera	Adults	Nishino and Sakai (1996), Honma et al. (2006), and Cadena-Castañeda et al. (2016)
				Phasmida	Adults	Godden (1972) and Farkas (2016)
				Mantodea	Adults	Edmunds (1972) and Lawrence (1992)
				Hemiptera	Adults	Holmes (1906), Villet (1999), and Kang et al. (2017)
				Neuroptera	Larvae	Sendova-Franks et al. (2020), Farji-Brener et al. (2022, 2024a, b), and Taylor et al. (2023)
				Coleoptera	Larvae	Miyatake et al. (2008) and Matsumura et al. (2016)

(continued)

Table 3.1 (continued)

Phylum	Subphylum	Class	Subclass	Order	Statge	References
					Adults	Fabre (1910), Duporte (1916), Bleich (1928), and Chemsak and Linsley (1970),
					Adults	Allen (1990), Oliver (1996), and Miyatake (2001a, b)
					Adults	Miyatake et al. (2004) and Hozumi and Miyatake (2005),
					Adults	Kuriwada et al. (2009) and Spencer and Richard (2013)
					Adults	Ritter et al. (2016) and Konishi et al. (2020)
					Adults	Okai and Sugiura (2023)
				Mecoptera	Larvae	Ishihara et al. (2024)
				Thysanoptera	Unknown	Sabelis and van Rijn (2002)
				Diptera	Adults	Gowda et al. (2023) and Sato and Takahashi (2025)
				Lepidoptera	Larave	Tojo (1991)

					Adults	Balletto et al. (1987), Dudley (1989), Larsen (1991), and Shreeve et al. (2006)
				Hymenoptera	Adults	Hölldobler and Wilson (1990) and van Veen et al. (1999),
					Adults	King and Leaich (2006), Cassill et al. (2008), and Neves and Pie (2018)
					Adults	Amemiya and Sasakawa (2021) and Fujioka et al. (2024)
					Adults (worker and queen)	Cardoso et al. (2024)
Chordata	Vertebrata	Actinoptergii	Pisces	Cichliformes	Adults	Mckaye (1981) and Barreto et al. (2003)
				Perciformes	Adults	Radabaugh (1989) and Gibran (2004)
				Gasterosteiformes	Adults	Messler et al. (2007)
				Cyprinodontiformes	Adults	Templeton and Shriner (2004) and Bass and Gerlai (2008)
				Labriformes	Juveniles	Resende et al. (2025)

(continued)

Table 3.1 (continued)

Phylum	Subphylum	Class	Subclass	Order	Statge	References
				Syngnathiformes	Adults	Freret-Meurer et al. (2016)
		Chondrichthyes	Elamobranchii	Hexanchiformes	Adults	Henningsen (1994)
				Squaliformes	Adults	Kreidl (1916), Mangold (1920), Lissman (1946), and Whitman et al. (1986)
				Rhinopristiformes	Adults	Henningsen (1994)
				Orectolobiformes	Adults	Brunnschweiler and Pratt (2008), Williamson et al. (2018), and Mukharror et al. (2020)
				Lamniformes	Adults	Pyle et al. (1999) and De Swaef et al. (2020)
				Carcharhiniformes	Adults	Gruber and Keyes (1981), Whitman et al. (1986), Watsky and Gruber (1990), and Davie et al. (1993)
					Adults	Henningsen (1994), Holland et al. (1999), and Brooks et al. (2011)
				Rajiformes	Adults	Schaefer (1921) and Henningsen (1994)

				Myliobatiformes	Asults	Henningsen (1994)
		Amphibia		Anura	Adults	Gally et al. (2012), Ramalho et al. (2019) and Toledo et al. (2010)
				Urodela	Larvae	Schulte and Caspers (2024)
		Reptilia		Squamata	Adults (Serpentes)	Gehlbach (1970), Burghardt and Greene (1988), Gerald (2008), and Marques et al. (2013)
					Adults (Serpentes)	Gregory and Gregory (2006) and Patel et al. (2016)
					Adults (Lacertilia)	Santos et al. (2010), Muscat et al. (2016), and Costa-Campos and Anaissi (2020)
				Testudines	Juveniles	Rusli et al. (2016)
		Aves		Accipitriformes	Adults	Vogel (1950)
				Galliformes	Adults	Gallup (1973) and Thompson et al. (1981)
				Anseriformes	Adults	Sargeant and Eberhardt (1975)

(continued)

Table 3.1 (continued)

Phylum	Subphylum	Class	Subclass	Order	Statge	References
		Mammalia		Lagomorpha	Adults	Giannico et al. (2014)
				Artiodactyla	Adults	Moore and Amstey (1962)
				Cetatriodactyla	Juvenile	Lundgren and Moeller (2017)
				Didelphimorphia	Adults	Franq (1969)
				Primates	Adults	Hennig (1978), Abrams et al. (2009), and Volchan et al. (2011)

3.1 Class Gastropoda

3.1.1 Order Neogastropoda

In the conch shell species *Pachyoliva semistriata*, predator-induced immobility has been observed in response to a carnivorous molluscan predator (Ulate Gómez et al. 2024). The authors describe this behavior as a novel form of "post-contact freezing," highlighting its occurrence following direct physical interaction with the predator.

3.2 Class Arachnida

3.2.1 Order Scorpiones

Thanatosis (tonic immobility) has been documented in *Tityus pusillus* (Lira et al. 2020), with adults exhibiting significantly longer durations of immobility compared to juveniles, and with longer durations occurring during the daytime relative to nighttime. Furthermore, death-feigning behavior has also been reported in *Thanatityus ocelote* and *Ananteris platnicki* (Triana et al. 2022).

3.2.2 Order Opiliones

Thanatosis and freezing behavior have been observed in the harvestman species *Mischonyx cuspidatus* (Segovia et al. 2019a). Comparative analysis of the duration of death-feigning behavior between starved and fed individuals revealed no significant differences. In a subsequent study, sex-based comparisons in the same species showed no significant variation in either the frequency or duration of death-feigning behavior between males and females (Segovia et al. 2019b).

3.3 Class Crustacea

3.3.1 Order Diplostraca

Yamada and Urabe (2021) conducted the first study investigating death-feigning behavior in cladocerans (*Daphnia*). They examined avoidance strategies against predation by dragonfly larvae using two *Daphnia* species, *Chydorus sphaericus* and *Oxyurella tenuicaudis*. Analysis of 1099 behavioral events revealed that upon encountering a predator, 38% of individuals swam away while 62% ceased movement and exhibited death-feigning behavior. The study demonstrated that this immobility significantly reduces predation risk from dragonfly larvae. The duration of death feigning varied considerably in both species, ranging from instantaneous (less than 0 s) to up to 200 s, with most immobilizations lasting less than

50 seconds. This work represents a valuable contribution to predator–prey research by clearly identifying predator species and empirically showing that death feigning effectively aids in predation avoidance (Yamada and Urabe 2021).

3.3.2 Order Isopoda

According to Hoagland (1927), Crozier (1923) conducted pioneering experiments on the curly woodlouse, *Cylisticus convexus* (an isopod species), examining the duration of tonic immobility—also referred to as death feigning—as a function of ambient temperature. To the best of current knowledge, this constitutes the earliest documented case of death-feigning behavior in isopods. Subsequently, Saxena (1957) reported that woodlice (*Armadillidium vulgare*) exhibited more pronounced tonic immobility at lower temperatures and under reduced lighting conditions.

More recently, Quadros et al. (2012) reported two key findings in several terrestrial isopod species of the genus Balloniscus: (1) the duration of death-feigning behavior differed significantly among species and (2) smaller individuals exhibited greater responsiveness to external stimuli during tonic immobility compared to larger individuals. In *Balloniscus sellowii*, no sex-related differences were observed.

In Japan, the invasive yet popular cockroach beetle species *Armadillidium vulgare* is well-known among children for its defensive curling behavior, commonly referred to as "rolling." This behavior is also classified as tonic immobility or death-feigning (Moriyama 2004). Adult females carrying eggs in their abdomen have been shown to be unable to perform this defensive response, which is presumed to represent a reproductive cost during incubation (Suzuki and Futami 2017). Recently, Igarashi and Wada (2024) reported that the marine isopod *Cleantiella isopus* exhibits tonic immobility characterized by freezing in a distinctive pose. This immobility is considered an adaptive defensive tactic in response to predatory stimuli from fish. The authors also documented sex differences in the frequency and duration of death feigning, with a negative correlation between body size and death-feigning duration observed in males.

Thanatosis has also been reported in the freshwater crab *Trichodactylus panoplus*, where the duration of thanatosis decreases with increasing body size (Scarton et al. 2009).

3.3.3 Order Decapoda

O'Brien and Dunlap (1975) reported that blue crabs (*Callinectes sapidus*) exhibited significantly shorter durations of tonic immobility when immobilized on a sandy substrate compared to those placed on a hard surface. In the spider crab genus *Chasmagnathus*, a distinct defensive behavior has been described wherein individuals adopt a freezing posture in response to an approaching object—experimentally induced by bringing a screen close to the animal (Pereyra et al. 1999). This freezing response is characterized by the crab flattening itself against the substrate, extending the second and third walking legs laterally, and flexing the chelae in front of the oral region. The behavior was observed to occur more frequently during the winter season.

Thanatosis has also been documented in two freshwater crustaceans of the family Aeglidae (*Aegla longirostri* and *Aegla denticulata denticulata*), members of the infraorder Anomura (Coutinho et al. 2013). In these species, no photoperiodic variation in the expression of tonic immobility was observed; however, females exhibited longer durations of thanatosis than males.

Immobility behavior has further been observed in five species of fiddler crabs (*Uca* spp.), with reported immobility durations ranging from 45 to 171 s. This response has been interpreted as a predator-avoidance strategy (Bergey and Weis 2006). Additionally, death-feigning behavior has been reported in the semi-terrestrial crab *Ocypode quadrata* (Dakin and Wortham 2023).

To date, however, there appear to be no published reports documenting thanatosis or related immobility behaviors in decapod crustaceans such as prawns, shrimps, or lobsters.

3.3.4 Order Araneae

Death-feigning behavior has been extensively studied in the context of mating in spiders. This phenomenon is closely associated with sexual cannibalism, wherein the female may consume the male during or after copulation, particularly in species exhibiting pronounced sexual size dimorphism. In such cases, thanatosis is not employed as an antipredator response, but rather as a strategy to avoid intraspecific predation by the mating partner.

For example, in the web-building and sexually cannibalistic spider *Pisaura mirabilis*, males have been observed to escape female attacks by entering a state of immobility in the presence of the female (Bilde et al. 2006). Males that perform thanatosis are reported to have higher mating success and longer copulation durations compared to those that do not (Hansen et al. 2008).

Similarly, in the funnel-weaving spider *Aterigena aculeata*, females have been observed to feign death during courtship interactions, potentially reducing the likelihood of eliciting aggressive responses from males and thereby facilitating mating with preferred partners (Liu et al. 2024). Liu et al. also proposed that this form of female sexual catalepsy may function as a signal of receptivity, offering mutual benefits to both the signaller and the receiver.

By contrast, to date, there are no known reports of thanatosis in ground-dwelling jumping spiders (family Salticidae). This absence may be related to the lack of sexual cannibalism commonly observed in this group.

3.3.5 Order Acarina

Tonic immobility has previously been described in mites (Ebermann 1991; Yoder et al. 2009), but had not been directly quantified in ticks until the study by Oyen et al. (2021). While earlier literature referred to death-feigning behavior in ticks, formal measurements were lacking. Oyen et al. (2021) conducted the first experimental assessment of tonic immobility duration in two tick species, Dermacentor variabilis and *Rhipicephalus sanguineus* (Acari: Ixodidae). Their results showed that the expression of death-feigning behavior decreased as starvation persisted, with individuals displaying increasingly aggressive responses under prolonged food deprivation.

3.4 Class Diplopoda

Thanatosis was reported in the millipede species *Pseudoleptodesmus rubescens* in a 2023 study by Bouzan et al. (2023). This represents one of the few documented cases of death-feigning behavior in Diplopoda.

3.5 Class Chilopoda

A giant centipede species from Colombia has recently been reported to exhibit tonic immobility in response to physical stimulation, a behavior interpreted as predator avoidance (Vega-Román et al. 2024). Although the natural predators involved remain unidentified, this study constitutes the first documented case of thanatosis-like mimicry behavior in centipedes.

3.6 Class Insecta

3.6.1 Order Odonata

Death-feigning behavior has been documented in damselfly larvae (Gyssels and Stoks 2005). Their study demonstrated that thanatosis increased larval survival when exposed to both dragonfly and fish predators. Additionally, larvae originating from fish-containing ponds exhibited longer durations of thanatosis compared to those from fishless ponds. The results also suggest that predator-derived chemical cues influence the duration of death-feigning behavior.

Thanatosis has also been observed in adult dragonflies (Khelifa 2017a). The first documented case of death-feigning behavior in the large dragonfly *Aeshna juncea juncea* was reported by Rassim Khelifa of the University of Zurich. Khelifa conducted detailed observations of the mating behavior of this species, noting that multiple males frequently pursued a single female during flight. Females occasionally resisted males physically and were sometimes subject to interference from rival males during copulation. This behavior exemplifies sexual conflict, wherein persistent male mating attempts are countered by female resistance (Arnqvist and Rowe 2005).

Sexual conflict occurs when the reproductive interests of males and females diverge, such that behaviors advantageous to one sex's reproductive success impose costs on the other (Arnqvist and Rowe 2005). Typically, males maximize their fitness by mating with multiple females, whereas females generally require fewer mates to fertilize all of their eggs.

Females of many insect species, including dragonflies, possess a specialized organ called the seminal vesicle, which stores sperm received from males (Simmons 2001). Sperm can remain viable for extended periods within these storage structures; for instance, in certain ant queens, sperm viability can persist for decades (Gotoh et al. 2023). Consequently, females typically do not need to mate with multiple males to ensure fertilization of all their eggs. In fact, multiple matings may be detrimental to females, as they increase the risk of disease transmission and exposure to predation during copulation. This dynamic suggests that, contrary to some assumptions, there may be no direct conflict of interest between males and females regarding the frequency of mating (Arnqvist and Rowe 2005).

In dragonfly populations with male-biased sex ratios, females often attempt to lay eggs before, during, or even after mating. According to Khelifa (2017b), females are subjected to persistent harassment by males throughout these stages. Females typically evade males by flying away, rejecting advances, or engaging in physical struggles. Despite these efforts, males may forcibly seize females by grasping their heads and pinching their thoraxes using specialized claspers located on their abdomen. In response, females sometimes perform a sudden dive into the water while immobile, a behavior referred to as "pair hijacking," in which the female submerges with the male still attached.

Startled males typically fly away, allowing females to lay eggs after the males have departed. Khelifa (2017b) observed that in 14 of 16 cases, females escaped male grasping by falling into the water and feigning death. This represents an 80% evasion success rate, which is higher than any other avoidance tactic documented except thanatosis. For comparison, wing-flapping as an escape behavior had a success rate of 25.9%, while clinging to the substrate and attempting to dislodge males had a success rate of 57.7%, both significantly less effective than death feigning. When male harassment persists, females repeatedly crash into the water's surface and employ thanatosis, effectively avoiding further sexual harassment. These findings are detailed in Chapter 3 of Khelifa's (2017b) publication.

Wildermuth et al. (2019) reviewed death-feigning behavior in dragonflies within the context of sexual conflict. Their review compiles and discusses numerous published reports on "drop and stop" behaviors observed in dragonflies.

3.6.2 Order Plecoptera

Freezing and thanatosis have been observed in larvae of the stonefly *Pteronarcys dorsata* (Plecoptera) (Moore and Williams 1990). This predator-avoidance behavior appears to be predator-specific, occurring in response to fish predators but not to crayfish (Moore and Williams 1990). Additionally, these larvae are known to secrete a noxious fluid from leg joints or exhibit death-feigning behavior when threatened. The species is primarily nocturnal, with adults active during spring and summer, and represents the largest stonefly species in North America (Moore and Williams 1990).

3.6.3 Order Orthoptera

Death-feigning behavior in grasshoppers has been the focus of physiological investigations aimed at elucidating its functions and underlying mechanisms (Nishino and Sakai 1996, 2021; Sakai 2021a, b). Using the cricket species *Gryllus bimaculatus* as a model organism, Nishino and Sakai (1996) demonstrated that a light grip applied to the forelegs and pronotum induces a reflexive thanatosis response. During this state, the respiratory rate was significantly reduced while the heart rate approximately doubled relative to resting conditions. Insects in a state of deep rigor exhibited unresponsiveness to external stimuli but could be readily aroused by mechanical provocations such as poking. The immobile posture typically persisted for 2–4 min, although durations exceeding 20 min were occasionally observed, with spontaneous recovery. Catalepsy was induced when legs were passively extended during thanatosis.

The flexion reflex is elicited by stimulation of either the campaniform sensory organs on the foreleg and the femoral chordotonal organ (FCO), or the spinous sensory area of the anterior thoracic region. However, among these receptors, only the FCO was implicated in the induction of the immobility state. Central neural mechanisms within the brain were found to be essential for maintaining thanatosis. Under semi-natural conditions, tonic immobility was observed to occur spontaneously as crickets attempted to enter narrow crevices, reflecting species-specific behavior. Nishino and Sakai (1996) proposed that this tonic immobility may serve as an adaptive antipredator strategy, facilitating escape from predators such as reptiles and amphibians in the crickets' natural environment.

Subsequently, Nishino et al. (1999) demonstrated that the expulsion of the delayed exor unit is mechanistically essential for the maintenance of both thanatosis and catalepsy. They further elucidated distinct functional roles of this behavioral substrate within identified neuronal subgroups of the femoral chordotonal organ. This mechanism was also implicated in sustaining tonic immobility of the femoral–tibial joint across a wide range of joint angles during thanatosis and catalepsy. In a subsequent detailed neurophysiological study, Nishino (2004) investigated the relationship between neuronal activity and thanatosis using the same cricket species. Focusing on the mechanical and physiological properties of the tibial retractor muscles, the study revealed that during thanatosis, motoneurons are actively inhibited, resulting in a rigid state of the leg flexor muscles (Nishino 2021).

In 2006, thanatosis in grasshoppers was characterized as a novel form of escape or physical defense against predators (Honma et al. 2006; Honma 2021). Honma et al. observed the feeding behavior of the spiny grasshopper *Criotettix japonicus* in the field and identified frogs as the primary predators in wetland habitats. When swallowed by a predatory frog, *C. japonicus* exhibits a distinctive defensive posture: its spiny hind legs are raised vertically, while thoracic spines, present since adulthood, extend horizontally. The vertically erect legs become rigid, resembling the shape of a water chestnut (Trapa species), which inspired the Japanese common name "Hishi-seed grasshopper" ("Hishi" referring to the water chestnut plant). This stiffened, immobile posture prevents frogs from successfully swallowing the grasshopper. Importantly, this behavior is not exhibited in response to bird pecking; the grasshopper only assumes this rigid posture when inside a frog's mouth. Thus, the thanatosis serves as a specialized swallow-avoidance, physical defense strategy specifically against frogs.

Separately, Cadena-Castañeda et al. (2016) reported thanatosis behavior in katydid species of the *Steirodontini genera* from the Amazon rainforest.

3.6.4 Order Phasmida

Stick insects are generally characterized by low levels of movement, raising the question of whether death-feigning behavior occurs in this group. Nevertheless, thanatosis has been documented in stick insects. Predation experiments involving praying mantises demonstrated that thanatosis serves as an effective defensive strategy against mantis attacks in the species *Carausius morosus* (Reitze and Nentwig 1991).

Repeated stimulation reduces the duration of death-feigning behavior in the stick insect *Carausius morosus* (Godden 1972). Using this species, Godden conducted detailed analyses of the movement and angular changes of the tibia and leg during thanatosis, examining leg extension and contraction as well as the leg angle at the onset and termination of the immobility state. More recently, Büschges et al. (2021) reported that the neural network controlling the tibial extensor and flexor muscles at the lower leg joint generates the motor output underlying catalepsy in this species.

Farkas (2016) investigated death-feigning behavior in *Timema cristinae* and found that smaller individuals exhibited a higher propensity for thanatosis compared to larger individuals. Additionally, younger individuals displayed more frequent death-feigning behavior than older conspecifics. Given that stick insects undergo incomplete metamorphosis, further research into the relationship between body size, developmental stage, and thanatosis is warranted.

3.6.5 Order Mantodea

The earliest report of death-feigning behavior in praying mantises is likely that of Edmunds (1972). Lawrence (1992) described observations in which male mantids sometimes exhibit freezing behavior at a distance from females while moving away after copulation, a behavior that may function as a strategy to avoid postcopulatory cannibalism.

3.6.6 Order Thysanoptera

Death-feigning behavior has been described in the thrips of the family Aeolothripidae (Cudadaceae) by Sabelis and van Rijn (2002). However, aside from this report, there are no other documented instances of thanatosis in thrips, and the detailed characteristics of this behavior remain largely unknown.

3.6.7 Order Hemiptera

There have been experiments with adults of the water praying mantis *Ranatra chinensis* in which 10 individuals were subjected to death feigning (Holmes 1906). Holmes (1906) examined the response to temperature and compared death-feigning duration at high (30 °C) and low (10–14 °C) temperatures. The results showed that death feigning duration was shorter at high temperatures. Death-feigning duration was also shorter under light irradiation than in darkness.

In adults of the cicada species *Nyara thanatotica*, individuals respond to disturbance by flying a short distance to a low perch or by dropping to the ground, often lying on their backs among leaf litter and exposing the white ventral abdominal membrane (Villet 1999). Villet (1999) notes that no similar thanatotic behavior has been reported in other cicada species.

Death-feigning behavior has been reported in adults of the spotted lanternfly (*Lycorma delicatula*), which exhibit red, black, white, and flamboyantly colored wings (Kang et al. 2017). Kang et al. (2017) found that early in the outbreak season, individuals typically responded to stimuli by jumping away and fleeing, whereas significantly more death-feigning behavior was observed later in the outbreak season.

I have personally observed death-feigning behavior in adults of the bug *Acanthocoris sordidus* (Thunberg) (Miyatake, personal observation). Similarly, adults of the leaf-footed bug *Merocoris distinctus* (Coreidae), native to the Americas, exhibit death feigning when stimulated by flipping onto their backs with legs closed; upon arousal, they use their legs to right themselves and stand up (Peraza-Sánchez 2024). My observations suggest that the death-feigning behavioral sequence of *Acanthocoris sordidus* closely resembles that of *M. distinctus*.

3.6.8 Order Neuroptera

Thanatosis behavior has recently been increasingly reported in larvae of Neuroptera, specifically within the green lacewing family Chrysopidae and the antlion family Myrmeleontidae. Taylor et al. (2023) demonstrated that starvation of Chrysopidae larvae for more than 1 day led to an increased frequency of thanatosis, accompanied by a shorter duration of immobility. Furthermore, they estimated the heritability of the propensity to enter tonic immobility in response to stimulus intensity, reporting a broad-sense heritability of 0.502.

The larvae of *Euroleon nostras*, a species within the family Myrmeleontidae commonly known as antlions, are sand-dwelling ambush predators. Studies on thanatosis in this species have often assumed birds to be their primary natural predators. Sendova-Franks et al. (2020) and Franks et al. (2021) proposed redefining thanatosis as "post-contact immobility" (PCI). Their modeling studies on antlion predator–prey interactions concluded that extending the duration of death feigning beyond a certain point does not further increase mortality risk, as predators are likely to shift focus to other antlions within the same patch. The model suggested that the optimal duration of death feigning is highly unpredictable, enhancing survival by preventing predators from anticipating the prey's immobility time.

Furthermore, Farji-Brener et al. (2022) demonstrated that the distribution of immobility durations in antlions follows a Poisson distribution with $\lambda = 1$. They argued that random variation in the duration of immobility is optimal for predator avoidance because it prevents predators from predicting when the prey will resume movement. This finding reinforces the notion that unpredictable death feigning durations enhance survival. Additionally, Farji-Brener et al. (2022) reported that both intrinsic and extrinsic factors contributing to variability in post-contact immobility times are often linked to traits that are difficult for predators to anticipate.

Antlion larvae construct pitfall traps, known as antlions, which function by causing prey to fall into the sand (Farji-Brener et al. 2024a). In other words, they are classic sit-and-wait predators. Nest morphology has been examined in the antlion species *Myrmeleon inmaculatus*, with similar nest structures likely present in closely related species. Farji-Brener et al. (2024b) studied larvae of another antlion species, *Myrmeleon crudelis*, in the field and found that the duration of immobility following contact tended to be shorter when larval nests were in close proximity and stimulation was applied. Antlions build funnel-shaped nests spaced at certain intervals within fixed sandy areas in open habitats; however, death-feigning duration decreases as nest density (i.e., population density) increases. Furthermore, Farji-Brener et al. (2024b) reported significant differences in death-feigning duration between laboratory-reared and field-collected larvae, with lab-reared individuals exhibiting significantly shorter immobility times than their wild counterparts.

Franks et al. (2024) investigated the dynamics of immobility duration in antlion larvae and found that larvae vary in their response depending on the substrate: whether they walk and flee or dive into the sand. This variation characterizes the dynamics of immobility duration. They reported that postcontact immobility (PCI) in antlion larvae represents a disruptive event, after which movement and immobility gradually return to a pattern typical of intermittent locomotion, influenced by the availability of substrate suitable for burying and hiding.

Building on this, Farji-Brener et al. (2025) examined the trade-off between PCI and burying behavior in *Myrmeleon immaculatus*. In response to predatory birds, the larvae faced a decision-making scenario where they had to choose between diving into the substrate or remaining immobile. Their results showed that individuals exhibiting longer diving behavior into the sand displayed shorter durations of death feigning.

3.6.9 Order Coleoptera

Among all taxonomic groups, Coleoptera (beetles) are the most species-rich and have the greatest number of studies investigating death-feigning behavior. Fabre (1910) was the first to experimentally study death feigning using beetles. Subsequently, Bleich (1928) provided detailed descriptions of death-feigning behaviors across numerous beetle species. Immobility induced by vibrations in the order Coleoptera has been reviewed by Takanashi and Kojima (2021).

3.6.9.1 Family Tenebrionidae

Jean-Henri Casimir Fabre (1823–1915) described his observations of insect thanatosis in a chapter entitled "Death Feigning" in Fabre's Book of Insects, published in 1900. Fabre placed an adult beetle, *Scarites aterrimus* (family Tenebrionidae), on his desk at home, stimulated it to induce death-feigning behavior, and observed the latency until the beetle resumed normal activity. He was particularly interested in the duration of thanatosis and conducted experiments varying the number of days the beetle feigned death as well as ambient temperature. Despite these efforts, Fabre was unable to determine the factors governing the duration of death feigning.

Additionally, Fabre placed a small-bodied ground beetle, *Scarites aterrimus*, feigning death on a table and then introduced a larger ground beetle, *Scarites sulcatus*, into the vicinity. As the larger beetle approached, the smaller beetle suddenly resumed movement and escaped. Fabre commented on this observation with a somewhat resigned tone, writing, "This is the time to play dead." Over the course of many experiments, Fabre grew increasingly uncertain about the adaptive significance of death feigning in beetles. Reflecting on his work, he questioned, "Can I get any satisfactory conclusions out of this jumble of muddled stuff? I hope so." He closed the chapter titled "Death Feigning" in his insect book with these words.

The subsequent report on death-feigning behavior in beetles was authored by Bleich (1928), although it was published in German. The study by Bleich (1928) has been recently reviewed by Pflüger et al. (2021); readers are referred to that review for detailed information.

The family Tenebrionidae also includes *Tribolium* beetles, such as *T. castaneum*. As discussed in detail in Chaps. 4 and 5, *T. castaneum* represents one of the most extensively studied beetle species regarding death-feigning behavior. The first research on death feigning in *T. castaneum* was a pioneering study that examined geographic variation in this behavior among populations (Prohammer and Wade 1981). Prohammer and Wade (1981) measured the duration of death feigning in *T. castaneum* individuals sourced from Campanario and Andújar in Spain, Kyoto in Japan, Arkansas in the United States, and a laboratory strain known as Chicago Black. Their results showed that populations from Arkansas and Japan exhibited significantly longer durations of death feigning compared to those from other regions.

No significant differences in death-feigning duration were observed between sexes across the eight regional populations. Additionally, Prohammer and Wade (1981) investigated the heritability of death-feigning duration by crossing individuals from Japan, characterized by long durations of death feigning, with those from Campanario (Spain), which exhibited shorter durations. The offspring displayed intermediate death-feigning durations relative to their parents, indicating genetic variation in this trait. Further diallel cross experiments estimated both paternal and maternal effects contributing to the genetic variance (Prohammer and Wade 1981).

High repeatability of death-feigning behavior and its association with resting metabolic rate have also been examined in the yellow mealworm, *Tenebrio molitor* (Krams et al. 2014).

Using the prey–predator system of *T. castaneum* and the Adanson jumping spider, Miyatake et al. (2004) were the first to demonstrate the effectiveness of death-feigning behavior in predator avoidance, employing robust data sets and rigorous statistical analyses. In their experiments, death-feigning duration was subjected to divergent artificial selection, resulting in the establishment of two populations: one exhibiting long death-feigning durations (L strains) and another with short durations (S strains). Comparative studies between these strains have since facilitated a wide range of behavioral, physiological, genetic, and molecular investigations (see Chap. 5).

A brief overview of the study using *T. castaneum* is provided here, with a more detailed description in Chaps. 4 and 5. Through 10 generations of artificial divergent selection, populations exhibiting long (L strains) and short (S strains) death-feigning durations were established. Realized heritability estimates confirmed the presence of genetic variation in death-feigning duration (Miyatake et al. 2004) (Fig. 3.1).

When predation rates were compared using Adanson jumping spiders as model predators, beetles from the L strains exhibited significantly lower predation rates compared to those from the S strains (Miyatake et al. 2004). However, males of the L strains, despite their superior predator avoidance, demonstrated reduced mating success relative to males from the S strains due to slower locomotor activity. This finding indicated a genetic trade-off between predation avoidance and mating ability (Nakayama and Miyatake 2009b) (Fig. 3.2).

Nishi et al. (2010) compared the levels of biogenic amines—including octopamine, tyramine, dopamine, and serotonin—in the brains of *T. castaneum* from different strains. Gas chromatography analysis revealed that the L strain exhibited lower concentrations of dopamine, tyramine, and octopamine compared to the S strain (Miyatake et al. 2008). When these biogenic amines were injected into the beetles, a concentration-dependent effect on death-feigning duration was observed,

Fig. 3.1 Adults of *Tribolium castaneum*. The left panel shows an individual exhibiting death feigning, characterized by a distinctive posture in which the beetle lies on its dorsum with all six legs held close to the body and remains motionless. The right panel shows an individual not exhibiting death feigning

Fig. 3.2 An adult *Tribolium castaneum* exhibiting death feigning in the presence of a female of Adanson jumping spider (*Hasarius adansoni*)

with dopamine showing the most pronounced impact: higher concentrations of injected dopamine significantly shortened the duration of death feigning (Nishi et al. 2010). Moreover, both oral administration and injection of caffeine, a dopamine activator, significantly reduced the death-feigning duration in L strain individuals (Nishi et al. 2010).

Stress tolerance has also been compared between the L and S strains. The L strains exhibited lower tolerance to heat shock and oscillatory stress regimes compared to the S strains. Additionally, expression levels of several stress-related genes—including catalase (*CAT*), heat shock proteins (*Hsps*), antioxidant enzymes, and growth-blocking peptide (*GBP*) genes—differed significantly between the strains (Kiyotake et al. 2014). These findings suggest that the prolonged death-feigning behavior in L strains is associated with physiological stress, indicating that death feigning incurs a measurable physiological cost.

Separately, Matsumura et al. (2016) performed divergent artificial selection on *T. castaneum* based on walking activity measured through trajectory tracking. They found that strains exhibiting higher locomotor activity showed a lower frequency of death-feigning behavior and shorter durations of thanatosis.

Artificial selection for death-feigning duration was continued for several dozen generations, resulting in an increased selection response. When selection pressure was subsequently relaxed, the trait gradually reverted toward baseline levels (Matsumura and Miyatake 2018). This pattern further supports the hypothesis that prolonged death feigning incurs a physiological cost (Table 3.2).

The first study investigating factors responsible for arousal from death feigning using artificially selected strains was conducted in *T. castaneum*. Specifically, the threshold of vibratory stimulus intensity required to elicit arousal from thanatosis was examined using the long-duration death-feigning strain (L strain) (Miyatake et al. 2019). The results showed that adult *T. castaneum* individuals from the L strain could be aroused from death feigning by vertical vibrations of approximately 0.2 mm applied to the substrate (Petri dish) on which they were placed (Miyatake et al. 2019).

Additionally, crosses between the short-duration (S) and long-duration (L) death-feigning strains, as well as F_2 segregation analyses derived from these crosses, revealed a genetic basis for the depth of death feigning—that is, the sensitivity to vibratory stimuli during thanatosis (Miyatake et al. 2019). The effect of aggregation pheromones as a stimulus for arousal from death feigning in *T. castaneum* was also investigated. It was demonstrated that adult beetles of the L strains, which exhibit prolonged death feigning, were aroused more rapidly when exposed to synthetic aggregation pheromones originally released by males (Ishikawa et al. 2024).

Molecular analyses were also conducted to elucidate the genetic basis of death-feigning duration between strains. Specifically, transcriptome analysis was performed by comparing RNA sequences from selected strains with reference sequences registered in gene banks for *T. castaneum* as a model organism (Uchiyama et al. 2019). The results identified 286 genes in the S strain and 232 genes in the L strain that showed differential expression relative to the reference genome data.

Table 3.2 Direct and correlated responses to selections for short (S) or long (L)durations of death feigning in beetles

Response	Traits	*T. castaneum*	*T. confusum*	*T. freemani*	*C. chinensis*
Direct	Duration of death feigning	S < L	S < L	S < L	S < L
Correlated	Locomotor activity	S > L	S > L	S > L	S > L
Correlated	Avoiding ability from predators	S < L	–	S < L	–
Correlated	Male mating ability	S > L	–	–	–
Correlated	Flight ability	–	–	–	S > L
Correlated	Stress tolerance	S > L	–	–	–
Correlated	Dopamine expression	S > L	S > L	–	–
Correlated	Dopamine related gene variation	S < L	–	–	–

Notably, many of these genes exhibited variation related to dopamine pathways within the tyrosine metabolism system (Uchiyama et al. 2019). These findings indicate that death-feigning duration is closely associated with genetic variation in dopamine-related genes and their associated networks, with multiple genes implicated in regulating this behavioral trait (Uchiyama et al. 2019).

Subsequently, whole-genome resequencing analysis was conducted, revealing genomic associations related to variation in death-feigning duration between strains. Significant genetic variation was identified in pathways including tryptophan metabolism, xenobiotic metabolism by cytochrome P450, longevity regulation, and circadian rhythm, with greater variation observed in strains exhibiting prolonged death feigning (Tanaka et al. 2021). This study suggests that multiple metabolic pathways and associated genes, beyond the tyrosine metabolism system (including dopamine), identified in earlier research, may form a complex network underlying the decision-making processes involved in death feigning as an antipredator behavioral strategy (Tanaka et al. 2021).

These data suggest that multiple genes contribute to the duration of death feigning. Supporting this, phenotypic segregation analyses of F_1 and F_2 generations from crossbreeding experiments between strains revealed a genetic basis for death-feigning duration. The shorter duration phenotype was found to be dominant, and the trait was characterized as polygenic (Matsumura and Miyatake 2022).

In the field, populations of *T. castaneum* that have coexisted for over 10 years with the predator assassin bug *Amphibolus venator* (Klug) [Reduviidae] (Konishi et al. 2020) were studied. A comparison of death-feigning durations between populations cohabiting with predators and those without such exposure revealed significantly longer durations in the predator-exposed populations. These findings suggest that cohabitation with predators has driven the evolution of prolonged death-feigning behavior (Konishi et al. 2020).

Recently, Miura et al. (2025) conducted an experimental evolution study comparing *T. castaneum* populations cohabitated with the predator *Amphibolus venator* and those reared without predators over multiple generations. Their results demonstrated that after only five generations of cohabitation, the duration of death feigning in *T. castaneum* was significantly prolonged compared to non-cohabitated populations. In contrast, the frequency of death feigning and walking activity remained unchanged after five generations. These findings support the hypothesis that death feigning in *T. castaneum* evolves under natural predator selection pressure in the field. Notably, *T. castaneum* is distributed throughout Japan, except in the northernmost region of Hokkaido Island.

Matsumura and Miyatake (2023a) collected *T.castaneum* populations from various latitudes across Japan and conducted interregional comparisons of death-feigning behavior. Their results indicated a positive correlation between latitude and death-feigning traits: populations from higher latitudes exhibited a higher frequency of death-feigning individuals and longer durations of death feigning (Matsumura and Miyatake 2023a).

Matsumura et al. (2025) also investigated the relationship between population density and death-feigning behavior. He found that the duration of death feigning significantly increased in individuals kept in isolation compared to their pretreatment state. Moreover, individuals cohabiting with S strains exhibited significantly shorter death-feigning durations after cohabitation, whereas the presence of L strains did not affect synchronization of activity among individuals (Matsumura 2025).

Additionally, artificial divergent selection experiments targeting high and low locomotor activity were conducted in *T.castaneum* (Matsumura and Miyatake 2015, 2023b). These studies demonstrated that strains with higher locomotor activity showed a lower frequency and shorter duration of death feigning compared to strains with lower locomotor activity, providing further evidence for the dopamine-mediated death-feigning syndrome (Matsumura and Miyatake 2022; Matsumura and Miyatake 2023b).

Walking patterns were analyzed using machine learning techniques with DeepHL applied to locomotion data collected by the ANTAM system (Nagaya et al. 2017) from artificially selected *T. castaneum* strains exhibiting long (L) and short (S) durations of death feigning (Maekawa et al. 2020, 2021). The analysis revealed that individuals from the S strain changed their walking angles more frequently than those from the L strain. Specifically, during angle changes, S strain beetles tended to slow down and execute turns, whereas L strain beetles maintained a straight walking trajectory without significant angular adjustments (Maekawa et al. 2020, 2021).

Predation avoidance behavior of *T. castaneum* against *Amphibolus venator*, a specialized predator of stored-product pests, has been further investigated. When attacked by *A. venator*, *T. castaneum* was observed to adopt tonic immobility by remaining stationary rather than the classical posture of lying on its back and feigning death (Asakura et al. 2022). Interestingly, in the previously described cohabitation experiment (Miura et al. 2025), leg contraction and death-feigning behavior evolved under artificial stimulation while the beetles were in the supine position. These findings suggest a complex evolutionary relationship between predator pressure and death-feigning behaviors, warranting further research.

Circadian rhythms, which exhibited molecular-level differences as reported by Uchiyama et al. (2019), were also compared between strains selected for differing durations of death feigning. The results revealed significant differences in the amplitude of the diurnal rhythm between strains, whereas no differences were observed in the circadian period length (τ) (Miyatake et al. 2022). Additionally, carbon dioxide (CO_2) emission rates were measured to compare metabolic rates between the long-duration (L) and short-duration (S) strains. No significant differences were found, indicating the absence of a genetic correlation between death-feigning duration and metabolic rate (Matsumura et al. 2025).

Tribolium confusum and *T. freemani,* two species closely related to *T. castaneum* (genus *Tribolium*), have also been subjected to divergent selection for death-feigning duration (Nakayama and Miyatake 2010; Konishi et al. 2020). When exposed to vibratory stimuli, the threshold intensity required to induce arousal from death feigning differed among these species. Among the three, *Tribolium freemani* required the highest intensity of vibration to be aroused (Ishihara et al. 2021).

The effect of aggregation pheromones on arousal from death feigning was investigated using *T. castaneum* as the study species. The results showed that both males and females emerged from their death-feigning state upon exposure to synthetic aggregation pheromones produced by male beetles (Ishikawa et al. 2024).

As mentioned earlier, artificial selection experiments on death-feigning duration have also been conducted in other species of the genus *Tribolium*, including T*ribolium confusum* (Nakayama and Miyatake 2010) and *T. freemani* (Konishi et al. 2020). More recently, Matsumura et al. (2025) investigated the effects of cohabitation with conspecific individuals on the duration of death feigning. The results showed that individuals housed alone exhibited a significant increase in death-feigning duration compared to pretreatment levels. Furthermore, individuals cohabitating with populations selected for short death-feigning duration (S populations) showed a significant decrease in their death-feigning duration after cohabitation, whereas cohabitation with populations selected for long death-feigning duration (L populations) did not affect synchronization of activity patterns. Matsumura et al. (2025) also provided the first evidence of synchronization of antipredator behavior in terms of death feigning, reporting that death-feigning durations tended to synchronize toward shorter durations among many individuals.

Death-feigning behavior has also been studied in the horned beetle *Gnatocerus cornutus* (Tenebrionidae). Matsumura et al. (2020) found that males with larger jaws were less likely to adopt death-feigning behavior in response to stimuli, although the duration of death feigning was not correlated with male body size. Subsequent experiments using strains of *G. cornutus* exhibiting differences in sex-specific antipredator behavior showed that males tend to evade predator attacks primarily through active movement, whereas females are more likely to evade by exhibiting immobility, synonymous with freezing behavior (Katsuki et al. 2024).

More recently, some studies have highlighted death feigning as an indicator behavioral trait of personality using *T. castaneum* (Li et al. 2025). Li et al. (2025) found that death-feigning duration, together with walking distance and shelter emergence time, were consistent individual traits. They also reported that the interaction between death-feigning duration and walking distance in males influenced their mating success, with either shy but active males or bold but inactive males achieving higher numbers of matings. This study suggests that personality traits, including death feigning, may be linked to mating and reproductive success. In other words, this research is notable as the first to incorporate death feigning into the study of animal personality traits.

The relationship between parasitism and death feigning has also been investigated. Nakao et al. (2025) demonstrated that the duration of death feigning is shorter in *Tribolium confusum* adults infected with *Wolbachia* compared to uninfected individuals.

3.6.9.2 Family Curculionidae

Death-feigning behavior in weevils has also been well studied. The first report was on the weevil species *Tychius picirostris* in 1915, which described that adult weevils exhibit death feigning by becoming immobile, contracting their legs, and attaching them to their abdomen (DuPorte 1916). Subsequently, research focused on the adult sweet potato weevil, *Cylas formicarius*. It was found that the occurrence of death feigning in *C. formicarius* is influenced by the behavior prior to stimulation (Miyatake 2001a). Adults walking on sweet potato stems and leaves exhibited lower frequency and shorter duration of death feigning, whereas those perched showed higher frequency and longer duration, with feeding adults displaying intermediate values. Additionally, in *C. formicarius*, the duration of death feigning significantly decreased under starvation conditions, eventually resulting in the inability to feign death when starvation became severe (Miyatake 2001b).

Cylas formicarius, a pest of sweet potatoes, is a target of Sterile Insect Technique (SIT) programs implemented on several Japanese islands for eradication purposes (Moriya and Miyatake 2001). In adult *C. formicarius*, mating experience significantly reduces the duration of death-feigning behavior, with fertilized females exhibiting a further pronounced reduction compared to virgin females (Kuriwada et al. 2009). Mass-reared *C. formicarius* adults maintained for 71 generations under SIT breeding conditions showed no significant difference in death-feigning duration relative to wild counterparts (Kuriwada et al. 2010a). Moreover, exposure to gamma irradiation at 200 Gy, a dose commonly employed in SIT protocols, did not alter death-feigning duration in the weevil (Kuriwada et al. 2010b). Investigations into the effect of age on death-feigning revealed that older adults (>35 days posteclosion) display markedly reduced durations of tonic immobility (Kuriwada et al. 2011).

Research on the death-feigning behavior of *Cylas formicarius* (weevil) has been conducted not only in Japan but also by research groups in the United States and China (Ouyang et al. 2020, 2023). Ouyang et al. (2020) investigated the trade-off between tonic immobility and mate-searching behavior, demonstrating that males engaged in mate search exhibited significantly shorter durations of death feigning compared to non-searching males, suggesting a negative correlation, or trade-off, between mate-search activity and tonic immobility in males. Furthermore, in *C. formicarius*, the duration of tonic immobility was significantly reduced during courtship and copulation (Ouyang et al. 2023). Subsequently, bidirectional artificial selection experiments targeting long and short durations of tonic immobility were performed to establish respective strains, which showed no significant differences in courtship frequency or fertilization success between the selected lines (Ouyang et al. 2023).

The effect of pheromones as a stimulus for arousal from death-feigning behavior in *Cylas formicarius* was investigated by Himuro and Miyatake (2024). Initially, they demonstrated that exposure to artificially synthesized sex pheromones significantly shortened the duration of death feigning in males, whereas females did not exhibit arousal from tonic immobility under the same conditions. Subsequently, raw females were placed inside mesh tea pots and exposed to females exhibiting death-feigning behavior. A notable finding from this experiment was that males were promptly aroused from death feigning only when a sexually mature, aged female was present in close proximity, but not when an immature female or a male was presented (Himuro and Miyatake 2024).

Death-feigning behavior has also been investigated in the snout weevil *Eucryptorrhynchus brandti*, a forest pest in China that infests various trees, including the tree of heaven (*Ailanthus altissima*). This species exhibits tonic immobility by retracting its legs and adopting a compact posture (Li et al. 2018). Li et al. (2018) demonstrated that starvation significantly reduced the duration of death feigning in adults. Unlike in *Cylas formicarius*, no sex-specific differences in starvation tolerance were observed in *E. brandti*; consequently, the negative correlation between starvation and the duration of death feigning was similar in both males and females.

Li and Wen (2021a) investigated the relationship between ambient temperature and the duration of death-feigning behavior in two weevil species, *Eucryptorrhynchus scrobiculatus* and *E. brandti*. In *E. scrobiculatus*, death-feigning duration was shorter at both low (15 °C) and high (32 °C) temperatures, whereas in *E. brandti*, it was prolonged at the lower temperature (15 °C). However, metabolic rate was not significantly correlated with the duration of death feigning in either species (Li and Wen 2021a). In a separate study, Li and Wen (2021b) reported that death-feigning duration decreased with age in both *E. scrobiculatus* and *E. brandti*. Furthermore, individuals of both species exhibited shorter durations of death feigning during the daytime compared to nighttime.

Li et al. (2023) further explored the neurogenetic mechanisms underlying death-feigning behavior in *Eucryptorrhynchus scrobiculatus* and identified mechanical stimulation as the primary trigger for arousal from tonic immobility. Their study revealed that the thoracoabdominal ganglion transmits signals to the brain via the ventral nerve cord, and that the brain modulates dopaminergic (DA) and serotonergic (5-HT) metabolic pathways by downregulating the expression of dopamine receptor gene dar2 and octopamine receptor genes oar1 and oab2. In a separate study, Wang et al. (2023) utilized high-speed videography to examine the detailed motor patterns associated with arousal from death feigning in *E. scrobiculatus* and *E. brandti*, specifically characterizing leg movements during the transition from immobility to active behavior.

In the vine weevil *Otiorhynchus sulcatus* (Curculionidae), the responsiveness of death-feigning adults to olfactory stimuli has been investigated using an olfactometer. The results suggest that exposure to specific odors can induce arousal from the death-feigning state (Nakamuta et al. 2005).

3.6.9.3 Family Coccinellidae

Death-feigning behavior is frequently observed in ladybird beetles (Coccinellidae) and was first documented in the pine ladybird *Exochomus quadripustulatus* by Allen (1990). Kudo and Hasegawa (2022) examined this behavior in *Harmonia axyridis*, a species exhibiting adult color polymorphism, and found that the frequency and duration of death feigning varied with color morph. Specifically, the red morph exhibited a longer duration of tonic immobility compared to the black morph, although this pattern was modulated by ambient temperature. No significant sex differences in death-feigning duration were observed. When tested at 25 °C, 20 °C, and 15 °C, the black morph (Type B) showed shorter death-feigning durations at lower temperatures, whereas the red morph (Type R) did not exhibit a clear temperature-dependent pattern. These results suggest a complex interaction between temperature and color polymorphism in the modulation of death-feigning behavior (Kudo and Hasegawa 2022).

Death-feigning behavior has also been reported in adult beetles of the family Endomychidae, a group within the ladybird beetle superfamily (Chang and Ren 2023).

3.6.9.4 Family Cerambycidae

In Cerambycidae, also known as the longicorn beetle, death-feigning behavior has been observed in *Hippopsis lemniscata* (Oliver 1996). This longicorn beetle has been reported to perform death feigning by freezing with its legs extended backward, rather than contracting them and attaching them to the body (Chemsak and Linsley 1970). In beetles, some, such as *T. castaneum*, assume an immobile pose with their legs contracted and attached to the body, while others, such as this longicorn beetle and *Cylas formicarius*, assume an immobile pose with their legs stretched out. Why two different poses exist for immobility is an unsolved mystery, and further research progress is desirable.

Although the author has personally observed death-feigning behavior in stag beetles since childhood, the first formal description of this behavior in English may have been in the species *Hoplogonus simsoni* (Spencer and Richard 2013). According to their report, *H. simsoni* exhibits immediate thanatosis in response to threat stimuli. Spencer and Richard (2013) also noted that such behavioral traits may contribute to the species' susceptibility to both predation and road mortality.

3.6.9.5 Family Lampyridae

Thanatosis has been reported in the larvae of the firefly *Lampyris noctiluca*, which exhibit this behavior by contracting their legs and overturning their bodies (Örstan and Novák 2022).

3.6.9.6 Family Chrysomelidae

Predation experiments involving praying mantises have demonstrated that thanatosis serves as an effective defense mechanism against predation in the leaf beetle *Cassida viridis* (Reitze and Nentwig 1991). The family Chrysomelidae, to which *C. viridis* belongs, also includes bean beetles (*Callosobruchus* spp.), which have been extensively studied with regard to their death-feigning behavior.

In the adzuki bean beetle, *Callosobruchus chinensis*, heavier adults—those reared at lower larval densities—exhibited longer durations of death feigning, with no significant sex differences observed in this trait (Hozumi and Miyatake 2005). Furthermore, Ohno and Miyatake (2007) conducted artificial divergent selection experiments on *C. chinensis*, targeting two traits: duration of death feigning and flight ability.

Artificial divergent selection for duration of death feigning over eight generations produced strains exhibiting a negative correlation between death-feigning duration and flight ability: strains selected for longer death feigning displayed reduced flight ability, whereas those selected for shorter death feigning showed enhanced flight ability (Ohno and Miyatake 2007). Conversely, artificial selection for high and low flight ability, measured as flight score, over 13 generations resulted in reciprocal changes in death-feigning duration, with strains possessing higher flight ability exhibiting shorter death-feigning durations, and vice versa (Ohno and Miyatake 2007). These findings indicate a negative genetic correlation and a trade-off between flight capacity and death-feigning duration in *Callosobruchus chinensis*.

Ohno and Miyatake (2007) further investigated the relationship between flight ability and death-feigning duration by analyzing 21 natural populations of *Callosobruchus chinensis* collected across East Asia. They found a significant negative correlation between death-feigning duration and flight ability in both males and females across these populations. This finding strongly supports the existence of a genetic trade-off between these two traits in natural environments, suggesting that natural selection acts on either flight ability or death-feigning duration within each population (Ohno and Miyatake 2007).

A negative relationship (negative phenotypic correlation) between ambient temperature and both the frequency and duration of death feigning was identified in two bean weevil species, *Callosobruchus maculatus* and *C. chinensis* (Miyatake et al. 2008).

In *Callosobruchus chinensis*, comparisons of life-history traits between lines subjected to artificial divergent selection for death-feigning duration revealed that the long-duration lines (L-lines) exhibited greater longevity, higher emergence rates, larger egg size, increased reproductive effort, and faster development compared to the short-duration lines (S-lines). Although L-lines tended to have longer lifespans, fertility did not differ significantly between L- and S-lines. Comparative analyses of life-history traits between lines subjected to artificial divergent selection for death-feigning duration revealed that long-duration lines (L-lines) exhibited increased longevity, higher emergence rates, larger egg size, greater reproductive effort, and a tendency toward accelerated development. Fecundity did not differ significantly between L- and short-duration lines (S-lines) (Nakayama and Miyatake 2009a). This suite of correlated traits has been termed the "death-feigning syndrome"(Nakayama and Miyatake 2010).

In the Mexican bean weevil, *Zabrotes subfasciatus*, a behavioral trade-off between fleeing and death feigning has been documented (Cardoso and Mendonça 2019).

3.6.9.7 Family Scarabaeidae

Thanatosis was characterized and analyzed as a behavioral personality trait in the dung beetle *Copris umbilicatus* (Natta et al. 2023). Natta et al. (2023) reported a negative correlation between thanatosis duration and locomotor activity, with individuals exhibiting shorter thanatosis durations displaying higher activity levels. No significant relationship was observed between the production of acoustic distress calls and thanatosis duration in this species. In the horned passalus beetle, *Odontotaenius disjunctus*, Davis et al. (2023) identified sex-specific effects of parasitism on stress-induced freezing behavior within a natural beetle–nematode system. They found that unparasitized females remained immobile for approximately twice as long as unparasitized males. From these findings, the authors inferred that heavily parasitized females tend to exhibit increased boldness, whereas heavily parasitized males are more likely to display timid behavior.

3.6.9.8 Family Dytiscidae

Okai and Sugiura (2023) investigated death-feigning behavior in 12 species of adult diving beetles by exposing individuals to artificial stimuli through dropping them into an indoor aquarium. They reported that eight species exhibited death feigning, which was interpreted as an antipredator strategy, whereas four species showed no such behavior. Additionally, their results indicated interspecific variation in responsiveness to different stimuli with respect to death-feigning behavior (Okai and Sugiura 2023).

3.6.9.9 Family Silphidae

Adults of the American burying beetle, *Nicrophorus americanus*, have also been reported to exhibit death-feigning behavior (Cavallaro et al. 2025).

3.6.9.10 Family Elateridae

Ritter et al. (2016) investigated antipredator behaviors in click beetles (genus *Agriotes*) and documented distinct forms of tonic immobility. Specifically, *Agriotes lineatus* exhibited pronounced tonic immobility, whereas A. obscurus and A. sordidus showed brief and inconsistent immobility responses. The authors measured the duration of tonic immobility, or death-feigning behavior, across five *Agriotes* species.

3.6.10 Order Mecoptera

The only study reporting death-feigning behavior in Mecoptera is by Ishihara et al. (2024). They observed thanatosis in the larvae of two scorpionfly species, *Panorpa japonica* and P. pryeri. Their results indicated that the form of the death-feigning posture changes with larval development: early instars (1st and 2nd) typically exhibit a rigid, stick-like immobility, whereas later instars (3rd and 4th) curl into a ball-like posture when feigning death. However, the adaptive significance of this ontogenetic change in posture remains unclear, as these larvae inhabit soil environments and their natural predators have yet to be identified. Additionally, the duration of thanatosis differs between the two species examined (Ishihara et al. 2024).

3.6.11 Order Diptera

To date, death-feigning behavior has not been widely reported in Diptera, although anecdotal observations suggested its occurrence. Notably, females of the robber fly *Efferia varipes* have been documented to exhibit death feigning as an anti-harassment strategy to avoid male mating attempts (Dennis and Lavigne 1976).

A recent study by Gowda et al. (2023) characterized the behavioral repertoire of *Drosophila* (likely *D. melanogaster*), including flipping and immobility responses upon capture, as well as traits related to activity, age, and neuronal activity. However, it remains challenging to conclusively determine whether such immobility or cessation of movement qualifies as thanatosis (death feigning). Thus, the interpretation of immobility as a distinct phenotypic behavioral trait analogous to thanatosis remains questionable. Namiki (2021) reviewed the brain controlling of freezing behavior at the neuron levels in *Drosophila melanogaster.*

In *Drosophila melanogaster*, freezing behavior triggered by predator attack has been examined across approximately 100 strains, revealing that isolated individuals exhibit pronounced freezing responses, whereas recovery from freezing occurs more rapidly when flies are in groups (Sato and Takahashi 2025). In contrast, personality studies of *Philornis downsi*, the avian vampire fly, assessed multiple behavioral traits but found no evidence of tonic immobility or death-feigning behavior (Common et al. 2025). This absence suggests that tonic immobility or death feigning may generally be uncommon or absent in dipteran flies.

3.6.12 Order Lepidoptera

The first report of thanatosis (death-feigning behavior) in Lepidoptera was in adults of a *Zygaena* moth species, which exhibit this behavior by crashing to the ground during flight and remaining immobile (Balletto et al. 1987). Subsequently, adults of the owl butterfly *Caligo illioneus* were observed to become immobile with their legs contracted against their bodies, a behavior described as tonic immobility (Dudley 1989).

Dudley (1989) suggested that "thanatosis may thus be a widespread antipredatory defense in tropical butterflies." Subsequently, Larsen (1991) observed thanatosis (death feigning) behavior in adults of two *Euploea* species (Lepidoptera), reporting durations ranging from less than 30 s to as long as 900 s, with the longest immobility typically between 121 and 180 s. Larsen proposed that this behavior functions as predator avoidance against birds, lizards, and snakes, and may represent a coevolutionary adaptation linking unpalatability and immobility. Additionally, Martinez and Ritzi (2020) reported on the duration of thanatosis in the salt marsh moth, *Estigmene acrea*.

Within Lepidoptera, immobility behavior has been reported in larvae of *Spodoptera litura*, which curl up and remain motionless (Tojo et al. 1985; Tojo 1991). Tojo (1991) also documented rapid arousal from immobility when larvae were reared at high densities, suggesting that physical contact among individuals may reduce the depth or duration of immobility. This finding represents a pioneering insight into the influence of tactile stimuli on immobility (see also Acheampong and Mitchell 1997; Miyatake et al. 2004). Additionally, Shreeve et al. (2006) described death-feigning behavior and conducted a phylogenetic analysis across Lepidopteran species. Reviews of moth freezing behaviors have been provided by Nakano (2021).

3.6.13 Order Hymenoptera

Tonic immobility has been studied in only a limited number of parasitic bee species and is recognized as an anti-predator defense behavior (King and Leaich 2006; Amemiya and Sasakawa 2021; Fujioka et al. 2024). Neves and Pie (2018) documented the behavioral repertoire of the sawfly *Perreyia flavipes*, which includes both death-feigning and sexual behaviors. In parasitoid wasps, immobility can be induced by gentle tactile stimulation of the antennae or abdomen (King and Leaich 2006; Amemiya and Sasakawa 2021).

King and Leaich (2006) reported that *Nasonia vitripennis* exhibited variations in both the frequency and duration of tonic immobility depending on background color. Similarly, *Heterospilus prosopidis*, another parasitoid wasp species, showed differences in immobility frequency related to background color (Amemiya and Sasakawa 2021). In H. prosopidis, immobility was more frequently observed in newly emerged individuals (0 days old) compared to older wasps (18 days old), with females displaying this behavior more often than males. Additionally, immobility occurred more frequently at lower temperatures (18 °C) than at 25 °C (Amemiya and Sasakawa 2021).

In *Heterospilus prosopidis*, variation in the frequency of immobility (pseudoparasitism) has been observed among local strains, with individuals reared from Hawaiian populations exhibiting lower frequencies compared to those from Arizona (Fujioka et al. 2024). Further sib analysis by Fujioka et al. (2024) revealed significant genetic variance in the frequency of pseudoparasitism, with both sire and dam effects detected. Additionally, Fujioka et al. (2024) reported that the frequency of pseudoparasitism varied with the age of the host, the adzuki bean weevil *Callosobruchus chinensis*, depending on the wasp strain. This finding aligns with the results of Amemiya and Sasakawa (2021), who found that pseudoparasitism frequency is higher at younger host ages and decreases with age across all strains.

This may be explained by the fact that newly emerged individuals at day 0 and day 1 possess limited flight ability, making immobility the most adaptive avoidance behavior at these early stages. However, since the natural predators of parasitic wasps in the field remain largely unknown and the existing experiments have been conducted under artificial, human-induced conditions, further research is needed to clarify the relationship between predation pressure and immobility behavior in natural environments. Additionally, death-feigning behavior has been documented in the stingless bee Melipona beecheii, where queens (gynes) reportedly feign death to evade attacks from worker bees (van Veen et al. 1999).

Within the order Hymenoptera, immobility (death-feigning behavior) has been reported not only in parasitic wasps but also in ants (Hölldobler and Wilson 1990). Cassill et al. (2008) documented that younger workers of the fire ant *Solenopsis invicta* exhibit immobility to avoid conflicts with competing ants. This tendency declines with age, as older workers are more likely to engage in aggressive interactions rather than remain motionless near conspecifics. Experimental evidence suggests that immobility enhances survival rates by a factor of four or more in young ants, likely due to their weaker exoskeletons (Cassill et al. 2008). Additionally, death-feigning behavior has been observed in *Polyrhachis femorata* (Hymenoptera: Formicidae) (Petit et al. 2023).

A recent literature review reported occurrences of death-feigning behavior in 13 ant species (Cardoso et al. 2024). The review also highlighted interspecific variation in the duration of immobility among these species. Additionally, death feigning has been documented in *Cataulacus granulatus*, an ant species native to southern China (Chen et al. 2025).

3.7 Class Actinoptergii

Haplochromis livingstoni, a cichlid species native to Lake Malawi, exhibits thanatosis (death-feigning behavior) during predation in two distinct manners (McKaye 1981): (1) the fish remains stationary with its ventral surface in contact with or near the substrate, subsequently rolling onto its side; or (2) the fish drops from the water column into a lateral recumbent posture (McKaye 1981). Thanatosis has also been documented in the cichlid *Parachromis friedrichsthalii* from Mexico, Central America, where this behavior is interpreted as a predatory strategy to ambush small prey fish, often referred to as a death-feigning hunting tactic (Tobler 2005).

Thanatosis has been documented in the seahorse *Hippocampus reidi* (Freret-Meurer et al. 2016). Observations conducted under both laboratory and field conditions led Freret-Meurer et al. (2016) to conclude that thanatosis serves as a secondary defense mechanism, activated following the failure or insufficiency of primary antipredator responses.

Additionally, a death-feigning strategy involving apparent simulation of illness or mortality to lure prey has been observed in the shallow-water predatory comb grouper, *Mycteroperca acutirostris* (Gibran 2004).

Adult individuals of the New Zealand endemic wrasse, *Notolabrus celidotus*, exhibit tonic immobility as an antipredator behavior, which is considered a secondary defense mechanism activated upon predator encounter (Resende et al. 2025). In the Nassau grouper, *Epinephelus striatus*, native to Puerto Rico, tonic immobility has been documented during handling procedures designed to observe behavioral responses (Tuohy et al. 2023).

3.8 Class Chondrichthyes

A comprehensive review on the adaptive significance of tonic immobility in elasmobranchs was published in 2023 (Páez et al. 2023). The review reported occurrences of tonic immobility in 21 species of sharks and rays spanning eight taxonomic orders (Páez et al. 2023). The high frequency of documented tonic immobility in sharks is partly attributed to unique research methodologies: many sharks are temporarily restrained and tagged during episodes of tonic immobility, facilitating behavioral studies and increasing observational records. Páez et al. (2023) proposed that tonic immobility in sharks and rays functions as an innate passive response to predation, plays a role in courtship and mating behaviors, or serves as a protective mechanism to mitigate the effects of overwhelming sensory stimuli, such as those experienced during tagging procedures. Regarding natural predators of sharks, Pyle et al. (1999) identified killer whales (*Orcinus orca*) as notable shark predators. It is plausible that tonic immobility in sharks evolved as an anti-predator strategy specifically to avoid predation by killer whales.

Tonic immobility behavior in certain shark species has been experimentally induced and characterized through physical restraint (Whitman et al. 1986; Watsky and Gruber 1990; Davie et al. 1993; Henningsen 1994). In a study investigating physiological stress responses associated with prolonged tonic immobility in the lemon shark (*Negaprion brevirostris*), Brooks et al. (2011) demonstrated that tonic immobility provoked significant alterations in multiple blood chemistry parameters and impaired respiratory efficiency. Despite these physiological challenges, compensatory mechanisms were observed that appeared to mitigate the detrimental effects of tonic immobility. Specifically, reductions in short-term ventilatory efficiency induced by tonic immobility were compensated by a marked increase in ventilation rate compared to control individuals, thereby enhancing overall gas exchange capacity. Furthermore, Brooks et al. (2011) proposed that structural and functional modifications within the gill and circulatory systems—commonly associated with stress responses—persisted throughout the tonic immobility period. These adaptations included recruitment of additional gill lamellae and vasodilation of branchial vasculature, both hypothesized to facilitate enhanced clearance of carbon dioxide.

In Western Australia, the Critically Endangered green sawfish (*Pristis zijsron*) has been reported to exhibit tonic immobility at the time of capture, during which individuals were marked for behavioral observation by researchers (Ingelbrecht et al. 2024). More recently, Gayford and Rummer (2025) published a comparative phylogenetic review on the evolution of tonic immobility behavior in Chondrichthyes, examining its functional significance and phylogenetic distribution. Their analysis included 29 species, with tonic immobility observed in over 50% of the examined taxa. The authors emphasized the need for further research to elucidate the evolutionary origins and adaptive functions of tonic immobility in this group. Yoshida (2021) also reviewed immobility behaviors in fish.

3.9 Amphibia

3.9.1 Class Anura

Predator avoidance behaviors in frogs have been comprehensively reviewed by Toledo et al. (2011). Death-feigning behavior, specifically, is addressed in a separate review by Toledo et al. (2010), which reported its occurrence in 5 out of 25 toxic frog species (20%) and in 35 out of 48 nontoxic species (73%). The authors also distinguished between thanatosis and a behavior termed "shrinking," noting that a higher proportion of individuals exhibited open eyes during thanatosis, whereas closed eyes were more common during shrinking. Gally et al. (2012) investigated the relationship between body size and death-feigning behavior in *Physalaemus kroyeri*, finding that smaller individuals were more likely to exhibit thanatosis. Death-feigning behavior has also been reported in toads of the genus *Incilius* (formerly Bufo) (Sánchez-Paniagua and Abarca 2016).

Since the review by Toledo et al. (2010), additional reports of death-feigning behavior in frogs have emerged. For instance, Ramalho et al. (2019) documented thanatosis in 12 frog species from the Amazon region. Klank et al. (2023) reported similar behavior in the floating rocket frog, *Silverstoneia flotator*, in Costa Rica. Furthermore, Folly et al. (2024) described thanatosis in *Pleurodema borellii* from Argentina. In my own observations, death-feigning behavior was also elicited in the Japanese tree frog, *Dryophytes japonicus*, in Japan by gently rubbing the abdominal region with a fingertip (Fig. 3.3).

Additional evidence supporting the association between tonic immobility and fear responses comes from a study on the Fijian ground frog (*Platymantis vitiana*) in response to predator exposure (Narayan et al. 2013). In this study, tonic immobility duration was used as an index of fearfulness following exposure to different stimuli. Frogs exhibited significantly longer tonic immobility durations after exposure to a predator toad compared to a conspecific or a neutral object (a ball), indicating that the intensity of the tonic immobility response correlates with the perceived threat level (Narayan et al. 2013).

Passos et al. (2017) reported differences in death-feigning behavior between wild and captive populations of golden mantella frogs (*Mantella aurantiaca*), suggesting that environmental context influences the expression of this behavior. More recently, Dittrich and Rödel (2023) demonstrated that female common frogs (*Rana temporaria*) employ three distinct avoidance behaviors to escape persistent courtship and forced mating attempts by males: rotational movements, release calls, and tonic immobility (i.e., death feigning). These findings suggest that, in this context, death feigning may function as an active strategy for sexual conflict resolution rather than as a passive antipredator response, as previously assumed.

Fig. 3.3 Death deigning of Japanese tree-frog, *Dryophytes japonicus*

Furthermore, Chiocchio et al. (2024) found that tonic immobility behavior in the yellow-bellied toad (*Bombina pachypus*) exhibits significant individual repeatability, indicating a potential behavioral trait with consistent individual variation. In *Dendropsophus branneri*, Velasco et al. (2025) identified death feigning as one component of a broader repertoire of antipredator strategies used by Neotropical anurans, including camouflage, limb retraction, and limb interweaving behaviors. Notably, photographs of startled frogs displaying diverse limb postures illustrate the behavioral plasticity of these responses.

3.9.2 Class Urodela

Death-feigning behavior (thanatosis) has long been recognized in salamanders, with individuals becoming immobile in response to threats (Brodie et al. 1974). More recently, Schulte and Caspers (2024) reported thanatosis in larval fire salamanders (*Salamandra salamandra*), demonstrating that even juvenile (larval) individuals exhibit this behavior as a predation avoidance strategy. In their study, larvae collected from both pond and stream environments were compared in terms of the duration of death-feigning behavior; however, no significant differences were found between the two habitat groups. Additionally, the duration of thanatosis was not correlated with larval body size (Schulte and Caspers 2024).

3.10 Reptilia

3.10.1 Class Squamata-Snake

There is a long-standing body of research on antipredator behaviors in snakes, including vigorous writhing and thrashing in response to predatory threats (comprehensively reviewed in Greene 1988). Among these strategies, death-feigning behavior (thanatosis) is particularly notable; it is characterized by immobility induced by predator handling, which may startle the predator and result in the release of the snake (Gehlbach 1970). A well-known example is the eastern hognose snake (*Heterodon platirhinos*), which, when threatened, exhibits a suite of behaviors including gaping, tongue protrusion, apnea (breath-holding), and complete immobility (Greene 1988). Experimental studies have often simulated predatory encounters—such as by having a human handler grasp and toss the snake, or by presenting an approaching owl—to reliably elicit death-feigning responses (Burghardt 1977).

Death-feigning behavior has been reported in aquatic snake species within the genus *Hydrodynastes* (Marques et al. 2013) and in *Erythrolamprus miliaris* (Muscat et al. 2016). Marques et al. (2013) found that five out of 13 individuals (38%) of *Hydrodynastes gigas* exhibited thanatosis when subjected to prolonged handling (approximately one minute). During thanatosis, individuals were observed with either a closed or open mouth. In a separate study, Gregory and Gregory (2006) investigated immobility responses in wild-caught garter snakes (*Thamnophis elegans*) under field conditions. While most individuals attempted to flee following release, a subset remained immobile—occasionally in a supine posture—for durations of up to 10 min.

Among various stimuli, the proximity of an owl following an initial freezing response, as well as direct human gaze, have both been shown to prolong the duration of death-feigning behavior in snakes, highlighting the complexity of factors influencing this response (Burghardt and Greene 1988). Mouth-gaping and thanatosis have been particularly well studied in the European grass snake (*Natrix natrix*). In a field study, Gregory et al. (2007) reported that 66% of captured *N. natrix* individuals exhibited death-feigning behavior. Notably, no significant relationship was found between body temperature and the duration of thanatosis in this species. Furthermore, in response to simulated predation by human grasping, juvenile (smaller-bodied) snakes were less likely to feign death and instead fled more readily compared to larger-bodied individuals (Gregory 2008).

In dice snakes (*Natrix tessellata*), some individuals exhibit a combination of defensive behaviors when confronted with a threat, including death feigning, autohemorrhaging (self-bleeding), and defecation (Bjelica and Golubović 2024). Interestingly, individuals that displayed both self-bleeding and defecation showed significantly shorter durations of death-feigning behavior. Based on these findings, Bjelica and Golubović (2024) emphasized the importance of studying the sequential expression of defensive behaviors, proposing that autohaemorrhage, defecation, and thanatosis may form part of an integrated antipredator behavioral repertoire.

Additional studies by Bjelica et al. (2025) revealed that pregnant *N. tessellata* do not feign death but instead exhibit tongue-flicking behavior in response to threats. Moreover, among color morphs, blotched (spotted) individuals displayed significantly longer durations of tonic immobility compared to green or melanistic (black-pigmented) individuals. Contrary to observations in the European grass snake (*Natrix natrix*), larger-bodied *N. tessellata* individuals exhibited longer durations of death feigning than smaller conspecifics.

Further research by Bjelica et al. (2025) found that juvenile (immature) individuals were less likely to feign death compared to adults. Their studies also explored the relationship between locomotor activity and death-feigning responses. Notably, populations of *N. tessellata* on mammal-poor islands exhibited slower escape speeds and a reduced frequency of death-feigning behavior, suggesting that environmental factors such as predator density and interspecific interactions influence the expression of thanatosis.

Death-feigning behavior has been documented in numerous additional snake species. For instance, the speckled kingsnake (*Lampropeltis holbrooki*) exhibits body coiling, tongue protrusion, and immobility as components of its thanatosis response (Muller et al. 2024). Similarly, death feigning has been observed in other members of the family Colubridae within the order Squamata, including *Malpolon monspessulanus* (Sannolo et al. 2014) and the ring-necked coffee snake (*Ninia diademata*) (Jones et al. 2024). Juvenile individuals of the twin-spotted wolf snake (*Lycodon jara*) have also been reported to perform thanatosis (Luitel and Devkova 2023). Notably, even the spectacled cobra (*Naja naja*) of the family Elapidae

(Cobraidae) has been observed exhibiting death-feigning behavior (Pandey et al. 2023). A comprehensive review of thanatosis in snakes, including observations across multiple species, was provided by Fuentes et al. (2021).

3.10.2 Class Squamata-Lizard

Death-feigning behavior has also been documented in several lizard species (Santos et al. 2010). Santos et al. (2010) observed thanatosis in *Liolaemus occipitalis* (family Liolaemidae), reporting that 75.6% of 86 individuals exhibited death feigning when stimulated, and that the duration of thanatosis increased with decreasing observer distance. Costa-Campos and Anaissi (2020) reported death-feigning behavior in eight lizard species inhabiting the Amazon rainforest. Hennig et al. (1976) found that Anolis carolinensis (Carolina anole) individuals immobilized near dense vegetation exhibited shorter durations of thanatosis compared to those immobilized in exposed areas without nearby cover. More recently, Rojas-Suárez et al. (2023) documented death feigning in the red-bellied ground lizard, *Stenocercus erythrogaster*, in Colombia. Alves and Cavalcanti (2024) reported thanatosis in *Colobosaura modesta* and suggested it may serve as an antipredator strategy against a range of potential predators, including snakes, lizards, birds, and amphibians. Additionally, O'Donnell (2024) observed death-feigning behavior in the Arizona night lizard, *Xantusia arizonae*.

Feigning death has also been documented in the gecko *Lygodactylus klugei* (Matias et al. 2024). Matias et al. (2024) additionally provided a comprehensive review of death-feigning behavior in lizards, reporting documented occurrences across eight lizard families. Death feigning has also been observed in *Lygosoma guentheri* and *L. punctata* (Patel et al. 2016). Notably, the terminology differs between taxa: in snakes, the terms tonic immobility or thanatosis are commonly used to describe this behavior, whereas in lizards, the term death feigning is more frequently employed.

3.10.3 Class Squamata-Testudines

In the green sea turtle (*Chelonia mydas*), newly emerged juvenile individuals have been observed to exhibit immobility lasting an average of 25 s (Rusli et al. 2016). This behavior has been characterized as tonic immobility and represents the first documented case of death-feigning behavior in turtles (Rusli et al. 2016).

3.11 Aves

The study of death-feigning behavior in birds has a long history and is supported by numerous publications. Vogel (1950) described a case observed in 1947 involving an adult turkey vulture (*Cathartes aura*) that froze on a tree branch and remained motionless for an extended period, occasionally moving only its eyes. In a seminal study, Gallup (1973) demonstrated that tonic immobility in chickens can be induced by electrical shocks or fear, further showing that the intensity of the stimulus positively correlates with the duration of tonic immobility. Gallup (1974) subsequently reviewed animal hypnosis, or tonic immobility, highlighting its utility as a quantifiable and measurable behavioral indicator.

Jones (1986) reviewed tonic immobility in domestic fowl, highlighting its association with physiological changes such as alterations in heart rate, respiration, and body temperature. He also discussed the role of genetic selection in modulating responsiveness to novel stimuli in domestic chicks, distinguishing lines bred for high and low reactivity. Furthermore, Jones (1986) reported a relationship between social status and tonic immobility duration, noting that dominant hens exhibited longer immobility responses compared to subordinate individuals.

In birds, the tonic immobility test is commonly employed as a behavioral indicator to assess welfare conditions in captive environments. This test is used to evaluate whether the rearing environment is conducive to bird well-being by measuring behaviors associated with fear and stress (e.g., in domestic hens: Archer and Mench 2017; Fiorilla et al. 2024; Özentürk 2024; Van Poucke et al. 2023, 2024; Ogundeji and Ayo 2025; and in turkey pied wrens: Ibrahim et al. 2024). Specifically, in hens, tonic immobility has been recognized as a reliable indicator of fear and stress (de Haas et al. 2013), with increased stress levels corresponding to longer durations of tonic immobility, which is used to evaluate the quality of the rearing environment (Skomorucha and Sosnówka-Czajka 2025). Comparisons of tonic immobility duration across different hen strains have found no significant differences (Akyol et al. 2024).

Hefnawy et al. (2024) investigated the effects of LED light color on behavior, including tonic immobility, using wild-type Muscovy ducks (*Cairina moschata*). They reported that ducks reared under blue LED lighting for 3 weeks exhibited longer durations of tonic immobility compared to those reared under white, red, or yellow LEDs. However, the influence of light color on tonic immobility varied with the duration of rearing; at 6 and 10 weeks, ducks exposed to yellow and red LEDs displayed higher levels of fear-induced tonic immobility than those under blue or white lighting (Hefnawy et al. 2024).

The Japanese quail (*Coturnix japonica*) utilizes tonic immobility behavior as an indicator of social relationships among individuals (Court et al. 2024). Classic studies of tonic immobility in quail have highlighted that predators such as cats preferentially target more active individuals, and that death feigning serves as a valid antipredator strategy employed as a last resort once prey are attacked (Thompson et al. 1981). An important foundational study by Sargeant and Eberhardt (1975) demonstrated, using ducks as a model, that death feigning increases survival against predators such as foxes; ducks that feign death after capture are often released by the predator, presumably due to the predator's reduced motivation to handle apparently unresponsive prey. Tonic immobility has also been observed in captive Amazon parrots (*Amazona aestiva* and *A. ventralis*) (Ramos et al. 2023).

Edelaar et al. (2012) conducted a study to assess intraindividual repeatability of tonic immobility (TI) responses in two avian species: the yellow-crowned bishop (*Euplectes afer*) and the tree sparrow (*Passer montanus*). Subjects were exposed to standardized predation risk stimuli via predator mounts positioned proximally to their home cages, alongside repeated manual handling to elicit TI. The results demonstrated significant individual consistency in TI duration across trials, with a robust negative correlation between TI duration and locomotor activity levels in the presence of predator stimuli. These findings support the interpretation of tonic immobility duration and activity level as behavioral traits indicative of individual personality axes, specifically boldness–shyness continua, thus evidencing a coherent behavioral syndrome that is stable across varying environmental contexts. In the rock pigeon (*Columba livia*), the relationship between tonic immobility and neural structures has been investigated using immunohistochemical techniques (Melleu et al. 2017).

In finch species, Suzuki et al. (2013) measured tonic immobility (TI) responses elicited by physical restraint as an indicator of fearfulness related to predator avoidance. Their study demonstrated that Bengalese finches exhibited significantly reduced tonic immobility durations compared to white-backed munias. Furthermore, TI responses were not influenced by sex, body weight, or rearing conditions.

3.12 Mammalia

Death-feigning behavior in opossums is widely recognized in English as "playing possum." Franq (1969) conducted detailed behavioral observations of thanatosis in captive opossums, systematically comparing death feigning with sleep. Key distinctions identified include: during sleep, eyes are closed, whereas during death feigning, eyes remain open; sleeping individuals rest in a dorsal-upward posture, while those feigning death lie with their lateral side upward. Additionally, in sleep, feet are tucked underneath the body, whereas in thanatosis, feet are visible with toes typically flexed or bent. Sleeping opossums show no response to sharp sounds, while those feigning death exhibit ear twitching in response to such stimuli. Furthermore, awakening from sleep involves a startle response and head orientation toward the stimulus source, whereas opossums in thanatosis show no reaction to poking or blowing, occasionally retracting their lips slightly. Remarkably, Franq (1969) documented that all six opossums studied engaged in death feigning for durations of at least 12 days, with one individual maintaining thanatosis for up to 60 days. Younger individuals were more prone to prolonged death feigning.

The American opossum (*Didelphis virginiana*) exhibits pronounced thanatotic behaviors characterized by contorted postures, open eyes, and tongues protruding from agape jaws. During thanatosis, these animals also commonly urinate, defecate, and secrete malodorous substances from their anal glands. Notably, *Didelphis opossums* possess a relatively low threshold for entering the thanatotic state, which may function as a form of deceptive camouflage employed preemptively before direct predator contact (Rogers and Simpson 2014).

Separately, Donatti and Leite-Panissi (2011) investigated the role of the corticotropin-releasing factor system in modulating emotional responses in guinea pigs (*Cavia porcellus*). Thanatosis behavior has been reported in the offspring of collared peccaries (*Pecari tajacu*), members of the peccary family within the suborder Suina, when confronted with predators such as gray foxes and bobcats (*Lynx rufus*) (Lundgren and Moeller 2017).

Moore and Amstey (1962) experimentally demonstrated that lambs are capable of exhibiting tonic immobility, whereas lambs placed in foster care at 1 year of age do not display this behavior, suggesting a significant maternal influence on its development. In a laboratory study, Ewell et al. (1981) showed that rabbits (*Oryctolagus cuniculus*) decreased the duration of death-feigning behavior as their proximity to a human increased or as they approached their home cage. Whishaw et al. (1978) also reported tonic immobility in some rabbit species.

Electrocardiographic studies in the rabbit (*Oryctolagus cuniculus*) have demonstrated significant alterations in heart rate during episodes of death feigning (Giannico et al. 2014). Giannico et al. (2014) confirmed and substantiated previous anecdotal observations by documenting pronounced reductions in heart rate alongside changes in systolic and diastolic cardiac function throughout the duration of tonic immobility.

Death feigning as tonic immobility has been found in two newborn males of domestic lambs (Skok et al. 2025). Skok et al. (2025) discussed that this reaction could be caused by thanatosis or by a genetic mutation that leads to a condition called myotonia congenita (Jungbluth et al. 2018), in which the goats freeze and fall over in response to a stressor. This might be an important suggestion for evolutionary views of tonic immobility in animals, because if disease caused by mutation and predator avoidance through death feigning have coevolved, it is possible that genes causing immobility-related diseases have been preserved over evolutionary timescales. This may provide insight into the relationship between instinct and death feigning, which will be discussed in Chap. 7.

Tonic immobility has also been reported in the squirrel monkey (*Saimiri sciureus*), with Hennig (1978) demonstrating that exposure to predator hawks increases vocalization and prolongs the duration of death-feigning behavior.

References

Abrams MP, Carleton RN, Taylor S, Asmundson GJG (2009) Human tonic immobility: Measurement and correlates. Depress Anxiety 26:550–556. https://doi.org/10.1002/da.20462

Acheampong A, Mitchell BK (1997) Quiescence in the Colorado potato beetle, *Leptinotarsa decemlineata*. Entomol Exp Appl 82:83–89

Akyol A, Şekeroğlu A, Tainika B, Şentürk YE, Duman M, Abacl SH, Gür FM (2024) Effects of strain and stocking density on performance, egg size distribution, egg quality, and welfare in laying hens housed in enriched cage system. Braz J Poult Sci 26:1–14

Allen AA (1990) Death-feigning in *Exochomus quadripustulatus* L. (Col Coccinelidae). Entomol Rec 102:23

Alves IG, Cavalcanti R (2024) Death-feigning behaviour in *Colobosaura modesta* (Reinhardt & Lütken, 1862). Herpetol Notes 17:521–522

Amemiya M, Sasakawa K (2021) Factors affecting thanatosis in the braconid parasitoid wasp *Heterospilus prosopidis*. Insects 12(1):48. https://doi.org/10.3390/insects12010048

Archer GS, Mench JA (2017) Exposing avian embryos to light affects post-hatch anti-predator fear responses. Appl Anim Behav Sci 186:80–84

Arnqvist G, Rowe L (2005) Sexual Conflict. Princeton University Press, p 348

Asakura M, Matsumura K, Ishihara R, Miyatake T (2022) Freezing or death feigning? Beetles selected for long death feigning showed different tactics against different predators. Ecol Evol 12:e8533

Balletto E, Arillo A, Mensi P, Cherchi MA (1987) Zygaenid moths, cyanide and Thanatosis. Ital J Zool 54:59–63

Barreto RE, Luchiari AC, Marcondes AL (2003) Ventilatory frequency indicates visual recognition of an allopatric predator in naïve *Nile tilapia*. Behav Processes 60:235–239

Bass SLS, Gerlai (2008) Zebrafish (*Danio rerio*) responds differentially to stimulus fish: the effects of sympatric and allopatric predators and harmless fish. Behav Brain Res 186:107–117

Bergey L, Weis JS (2006) Immobility in five species of fiddler crabs, genus *Uca*. J Crustac Biol 26:82–84

Bilde T, Tuni C, Elsayed R, Pekar S, Toft S (2006) Death feigning in the face of sexual cannibalism. Biol Lett 2:23–25

Bjelica V, Golubović A (2024) Synergistic effects of musking and autohaemorrhaging on the duration of death feigning in dice snakes (*Natrix tessellata*). Biol Lett 20:20240058. https://doi.org/10.1098/rsbl.2024.0058

Bjelica V, Marićić M, Anđelkovic M, Tomović L, Arsovski D, Golubović A (2025) The slithering dead: does locomotor performance affect post-capture death feigning in dice snakes (*Natrix tessellata*, Laurenti 1768)? Curr Zool 71:353–361

Bleich OE (1928) Thanatose und Hypnose bei Coleopteren. Zeitschrift für Morphologie und Ökologie der Tiere 10:1–61

Bouzan R, Ribeiro EA, Zanon MS, Brescovit AD, Iniesta LFM (2023) Back from the dead: a thanatosis report in *Pseudoleptodesmus rubescens* (Gervais, 1836) (Diplopoda: Polydesmida, Chelodesmidae). Entomol Commun 5:ec05036. https://doi.org/10.37486/2675-1305.ec05036

Brodie ED, Johnson JA, Dodd CK Jr (1974) Immobility as a defensive behavior in salamanders. Herpetologica 30:79–85

Brooks EJ, Sloman KA, Liss S, Hassan-Hassanein L, Danylchuk AJ, Cooke SJ, Mandelman JW, Skomal GB, Sims DW, Suski CD (2011) The stress physiology of extended duration tonic immobility in the juvenile lemon shark, *Negaprion brevirostris* (Poey, 1868). J Exp Mar Biol Ecol 409:351–360

Brunnschweiler JM, Pratt HL Jr (2008) Putative male–male agonistic ehavior in free-living zebra sharks, *Stegostoma fasciatum*. Open Fish Sci Jl 1:23–27

Burghardt GM (1977) Learning processes in reptiles. In: Gans C, Tinkle D (eds) Biology of the reptilia, vol 7. Academic Press, New York, pp 555–681

Burghardt GM, Greene HW (1988) Predator simulation and duration of death feigning in neonate hognose snakes. Anim Behav 36:1842–1844

Büschges A, Pflüger H, Bässler U (2021) Catalepsy and twig mimesis in insects and its neural control. In: Sakai M (ed) Death-feigning in insects, entomology monographs. Springer Nature Singapore Pte Ltd, Singapore, pp 135–143

Cadena-Castañeda OJ, Mendes DMDM, Alves-Oliveira JR (2016) A new genus of katydid from the Amazon rainforest (Orthoptera: Tettigoniidae; Phaneropterinae; Steirodontini): ninth contribution to the suprageneric organization of the Neotropical phaneropterines. Zootaxa 4150:493–500

Cardoso JCF, Mendonça JS (2019) The trade-off between fleeing and tonic immobility behaviors in an ectothermic animal. acta ethologica 22:129–134

Cardoso DC, Alves IC, Cristiano MP, Heinze J (2024) Death feigning in ants. Myrmecol News 34:45–56

Cassill DJ, Vo K, Becker B (2008) Young fire ant workers feign death and survive aggressive neighbors. Naturwissenschaften 95:617–624

Cavallaro M, Hladik M, McMurry SR, Hittson S, Boyles LK, Hoback WW (2025) Neonicotinoid exposure causes behavioral impairment and delayed mortality of the federally threatened American burying beetle, *Nicrophorus americanus*. PLoS One 20(1):e0314243. https://doi.org/10.1371/journal.pone.0314243

Chang L, Ren G (2023) New and little-known species of the genus *Cyclotoma* from China (Coleoptera: Endomychidae). Acta Entomol 63:177–185

Chemsak JA, Linsley EG (1970) Death-feigning among North American Cerambycidae (Coleoptera). Pan-Pacific Entomol 46:305–307

Chen Z, Cai W, Zeng Y (2025) Shape-shifting and rolling in an arboreal ant from southern China (*Cataulacus granulatus*). J Nat Hist 59:13–16

Chiocchio A, Martino G, Bisconti R, Carere C, Canestrelli D (2024) Shock or jump: deimatic behaviour is repeatable and olymorphic in the Apennine yellow-bellied toad, *Bombina pachypus*. Biol J Linn Soc 41:323–331

Common LK, Morales KA, Katsis AC, Kleindorfer S (2025) Personality in the parasitic avian vampire fly (*Philornis downsi*). Behaviour. https://doi.org/10.1163/1568539X-bja10317

Costa-Campos CE, Anaissi JC (2020) Death-feigning as defensive behaviour of eight lizard species of the Amazon rainforest. Herpetol Bull 152:26–27

Court L, Talbottier L, Lemarchand J, Cornilleau F, Pecnard E, Blache M, Balthazart J, Cornil CA, Keller M, Calandreau L, Pellissier L (2024) Exploring neuronal markers and early social environment influence in divergent quail lines selected for social motivation. Sci Rep 14:23554

Coutinho C, Ayres-Peres L, Araujo PB, Jara CG, Santos S (2013) Thanatosis in freshwater anomurans (Decapoda: Aeglidae). J Nat Hist 47:41–42

Crozier WJ (1923) Reflex immobility and the central nervous system. Exp Biol Med 21:55–56

Dakin M, Wortham JL (2023) Grooming and agonistic behaviors in the semi-terrestrial crabs *Ocypode quadrata* (Fabricius, 1787) and *Minuca* (*Uca*) *burgersi* (Holthuis, 1967) (Decapoda: Brachyura: Ocypodidae). J Crustacean Biol 43. https://doi.org/10.1093/jcbiol/ruad029

Davie PS, Franklin CE, Grigg GC (1993) Blood pressure and heart rate during tonic immobility in the black tipped reef shark. *Carcharhinus Melanoptera*. Fish Physiol Biochem 12:95–100

Davis AK, Ladd RRE, Shattuck A (2023) Sex-specific effects of a parasite on stress induced freezing behavior in a natural beetle nematode system. PLoS One 18(3):e0281149. https://doi.org/10.1371/journal.pone.0281149

de Haas EN, Kemp B, Bolhuis JE, Groothuis T, Rodenburg TB (2013) Fear, stress, and feather pecking in commercial white and brown laying hen parent-stock flocks and their relationships with production parameters. Poult Sci 92:2259–2269

Dennis DS, Lavigne RJ (1976) Ethology of *Efferia varipes* with comments on species coexistence (Diptera: Asilidae). J Kansas Entomol Soc 49:48–62

De Swaef E, Vercauteren M, Pieraets L, Declercq AM (2020) Fight, flight or freeze. Tonic immobility in sharks. Vlaams Diergeneeskd Tijdschr 89:243–250

Dittrich C, Rödel MO (2023) Drop dead! Female mate avoidance in an explosively breeding frog. R Soc Open Sci. https://doi.org/10.6084/m9.figshare.c.6858610

Donatti AF, Leite-Panissi CRA (2011) Activation of corticotropin-releasing factor receptors from the basolateral or central amygdala increases the tonic immobility response in Guinea pigs: an innate fear behavior. Behav Brain Res 225:23–30

Dudley R (1989) Thanatosis in the neotropical butterfly *Caligo illioneus* (Nymphalidae: Brassolinae). J Res Lepidoptera 28:125–126

DuPorte EM (1916) Death feigning reactions in *Tychius picirostris*. J Anim Behav 6:138–149

Ebermann E (1991) Thanatosis or feigning death in mites of the family Scutacaridae. In: The Acari: reproduction, development and life-history strategies. Springer, Dordrecht, pp 399–401

Edelaar P, Serrano D, Carrete M et al (2012) Tonic immobility is a measure of boldness toward predators: an application of Bayesian structural equation modelling. Behav Ecol 23:619–626

Edmunds M (1972) Defensive behavior in Ghanaian praying mantids. Zool J Linnean Soc 51:1–32

Ewell AH, Cullen JM, Woodruff ML (1981) Tonic immobility as a predator-defense in the rabbit (Oryctolagus cuniculus). Behav Neural Biol 31:483–489

Fabre JH (1910) Souvenirs Entomologiques. Delgrave Paris, Paris

Farji-Brener AG, Cajas A, Campos-Alvarado A, Castaño-Diaz M, Estévez-Haro M, Piñanez-Espejo YM (2022) Understanding unpredictability: factors influencing how long antlion larvae play dead as an antipredator behaviour. Anim Behav 194:199–204

Farji-Brener AG, Carrillo-Fajardo MY, Rodríguez-Malacara JT, Arroyo-Rodríguez V (2024a) Why are antlion larvae rare under the leaf litter? Testing the hypothesis of improper trap maintenance. Ecol Entomol 49:138–144

Farji-Brener AG, Abarca-Méndez A, Cubero-Morales L, López-Reyes K, Ramírez-Mosquera D, Escalante I (2024b) Neighbour density and post-contact immobility duration as antipredator behavior: antlion larvae do not fit the selfish prey hypothesis. Ethology 131:e13527

Farji-Brener AG, Enríquez M, Rosey D, Arroyo-Gerala P, Arroyo-Rodríguez (2025) What determines the antipredator strategy in antlion larvae? Burrowing ability decreases the duration of postcontact immobility. Ecol Entomol 50:115–119

Farkas TE (2016) Body size, not maladaptive gene flow, explains death-feigning behaviour in *Timema cristinae* stick insects. Evol Ecol 30:623–634

Fiorilla E, Ozella L, Sirri F, Zampiga M, Piscitelli R, Tarantola M, Ponzio P, Mugnai C (2024) Effects of housing systems on behaviour and welfare of autochthonous laying hens and a commercial hybrid. Appl Anim Behav Sci 274:106247

Folly H, Stazzonelli C, Assis CL (2024) Defensive behaviors in *Pleurodema borellii* (Anura: Leptodactylidae) from Northwest Argentina. Cuadernos de Herpetología 38:00–00

Franks NR, Worley A, Sendova-Franks AB (2021) Hide-and-seek strategies and post-contact immobility. Biol Lett 17:20200892

Franks NR, Worley A, Fortune GT, Goldstein RE, Sendva-Franks AB (2024) Seeking safety: movement dynamics after post-contact immobility. PLoS One 19(8):e0307370

Franq EN (1969) Behavioral aspects of feigned death in the opossum *Didelphis marsupialis*. Am Midl Nat 81:556–568

Freret-Meurer N, Fernandez TC, Lopes DA, Vaccani AC, Okada NB (2016) Thanatosis in the Brazilian seahorse *Hippocampus reidi* Ginsburg,1933 (Teleostei: Syngnathidae). Acta Ethol 20:81–84

Fuentes R, Castillo M, Belton E, Zambrano E, Quintero H, Batista A (2021) Dead snake! A strategy for survival: Thanatosis in some Panamanian snakes with a review of death-feigning in American snakes. Reptiles Amphibians 28:389–396

Fujioka H, Miyatake T, Ohkubo Y, Shimada M (2024) Hereditary nature of death-feigning frequency in a parasitoid wasp *Heterospilus prosopidis* using sib analysis. J Evol Biol 37:978–985

Gallup GG Jr (1973) Tonic immobility in chickens: is a stimulus that signals shock more aversive than the receipt of shock? Anim Learn Behav 1:228–232

Gallup GG Jr (1974) Animal hypnosis: factual status of a fictional concept. Psychol Bull 81:836–853

Gally M, Silvia ASFL, Zina J (2012) Death feigning in *Physalaemus kroyeri* (Reinhardt and Lütken, 1862) (Anura, Leiuperidae). Herpetol Notes 5:133–135

Gayford JH, Rummer JL (2025) Tonic immobility in cartilaginous fishes (Chondrichthyes): function, evolutionary history, and future directions. Rev Fish Biol Fisheries. https://doi.org/10.1007/s11160-025-09958-3

Gehlbach FR (1970) Death-feigning and erratic behavior in Leptotyphlopid, colubrid, and elapid snakes. Herpetologica 26:24–34

Gerald GW (2008) Feign versus flight: influences of temperature, body size and locomotor abilities on death feigning in neonate snakes. Anim Behav 75:647–654

Giannico AT, Lima L, Lange RR, Froes TR, Montiani-Ferreira F (2014) Proven cardiac changes during death-feigning (tonic immobility) in rabbits (*Oryctolagus cuniculus*). J Comp Physiol A 200:305–310

Gibran FZ (2004) Dying or illness feigning: an unreported feeding tactic of the comb grouper *Mycteroperca acutirostris* (Serranidae) from the Southwest Atlantic. Copeia 2004:403–405

Godden DH (1972) The innervation of the leg musculature and motor output during thanatosis in the stick insect *Carausius morosus* Br. J Comp Physiol 80:201–225

Gotoh A, Takeshima M, Mizutani K (2023) Near-anoxia induces immobilization and sustains viability of sperm stored in ant queens. Sci Rep 13:3029

Gowda SBM, Banu A, Salim S, Peker KA, Mohammad F (2023) Serotonin distinctly controls behavioral states in restrained and freely moving *Drosophila*. iScience. https://doi.org/10.1016/j.isci.2022.105886

Greene HW (1988) Antipredator mechanisms in reptiles. In: Biology of the reptilia. Vol. 16 (Ed. by C. Oans & R. B. Huey), Alan Liss, Inc, New York, 1–152

Gregory PT, Gregory LA (2006) Immobility and supination in garter snakes (*Thamnophis elegans*) following handling by human predators. J Comp Psychol 120:262–268

Gregory PT, Issac LA, Griffiths RA (2007) Death feigning by grass snakes (*Natrix natrix*) in response to handling by human "predators". J Comp Psychol 121(2):123–129

Gregory PT (2008) Bluffing and waiting: handling effects and post-release immobility in a death-feigning snake (*Natrix natrix*). Ethology 114:768–774

Gruber SH, Keyes RS (1981) Keeping sharks for research. In: Hawkings AD (ed) Aquarium systems. Academic Press, London, pp 373–402

Gyssels FGM, Stoks R (2005) Threat-sensitive responses to predator attacks in a damselfly. Ethology 111:411–423

Hansen LS, Gonzalez SF, Toft S, Bilde T (2008) Thanatosis as an adaptive male mating strategy in the nuptial gift–giving spider *Pisaura mirabilis*. Behav Ecol 19:546–551

Hefnawy E, Elgazzar E, Sabek A, El-laithy AS (2024) Effect of different LED light colors on welfare, performance, some behavioral patterns, and blood parameters of Muscovy ducks. BMC Vet Res 20:350

Hennig CW (1978) Tonic immobility in the squirrel monkey (*Saimiri sciureus*). Primates 19:333–342

Hennig CW, Dunlap WP, Gallup GG (1976) The effect of distance between predator and prey and the opportunity to escape on tonic immobility in "*Anolis carolinensis*". Psychol Rec 26:313

Henningsen AD (1994) Tonic immobility in 12 elasmobranchs: use as an aid in captive husbandry. Zool Biol 13:325–332

Himuro C, Miyatake T (2024) Mate search or predation avoidance? Sex pheromone interrupts death feigning of males in the sweet potato weevil *Cylas formicarius*. J Ethol 42:169–176

Hoagland H (1927) Quantitative aspects of tonic immobility in vertebrates. Proc Natl Acad Sci USA 13:838–843

Holland KN, Wetherbee BM, Lowe CG, Meyer CG (1999) Movements of tiger sharks (*Galeocerdo cuvier*) in coastal Hawaiian waters. Mar Biol 134:665–673

Hölldobler B, Wilson E (1990) The ants. The Belknap Press of Harvard University Press, Cambridge

Holmes JS (1906) Death-feigning in *Ranatra*. J Comp Neurol Psychol 16:200–216

Honma A (2021) The function of tonic immobility: review and prospectus. In: Sakai M (ed) Death-feigning in insects, entomology monographs. Springer Nature Singapore Pte Ltd, Singapore, pp 27–37

Honma A, Oku S, Nishida T (2006) Adaptive significance of death feigning posture as a specialized inducible defence against gape-limited predators. Proc Royal Soc B 273:1631–1636

Hozumi N, Miyatake T (2005) Body-size dependent difference in death-feigning behavior of adult *Callosobruchus chinensis*. J Insect Behav 18:557–566

Ibrahim RR, Azeem NMA, Emeash H, Abdelghany AK (2024) Performance, behavior, and welfare of Turkey poults reared under different housing conditions. J Adv Vet Res 14:30–36

Igarashi K, Wada S (2024) Tonic immobility in a marine isopod: the effects of body size, sex, and colour morph. J Mar Biol Assoc UK 104:e17. https://doi.org/10.1017/S0025315424000134

Ingelbrecht J, Lear KO, Phillips NM, Wueringer BE, Lymbery AJ, Norman BM, Morgan DL (2024) Kinship assessment and insights into reproductive behaviour of the critically endangered green sawfish *Pristis zijsron* in Western Australia. Aquat Conserv Mar Freshw Ecosyst 34:e4213

Ishihara R, Mastumura K, Jones JE, Yuhao J, Fujisawa R, Nagaya N, Miyatake T (2021) Arousal from death feigning by vibrational stimuli: comparison of *Tribolium* species. J Ethol 39:107–113

Ishihara R, Matsumura K, Miyatake T (2024) Death feigning in larvae of scorpionflies (Mecoptera: Panorpidae): frequency and postural changes based on larval instars. J Insect Behav 37:152–158

Ishikawa M, Matsumura K, Miyatake T (2024) Aggregation pheromone interrupts death feigning in the red flour beetle *Tribolium castaneum*. J Ethol 42:71–75

Jones RB (1986) The tonic immobility reaction of the domestic fowl: a review. Worlds Poult Sci J 42:82–96

Jones D, Kilburn V, Charette M (2024) Death-feigning behavior in the ring-necked coffee snake, *Ninia diademata* Baird and Girard 1853. Reptiles and Amphibians 31:e21459

Jungbluth H, Treves S, Zorzato F, Sarkozy A, Ochala J, Sewry C, Phadke R, Gautel M, Muntoni F (2018) Congenital myopathies: disorders of excitation–contraction coupling and muscle contraction. Nat Rev Neurol 14:151–167

Kang C, Moon H, Sherratt TN, Lee SI, Jablonski PG (2017) Multiple lines of anti-predator defence in the spotted lanternfly, *Lycorma delicatula* (Hemiptera: Fulgoridae). Biol J Linn Soc 120:115–124

Katsuki M, Kiyose K, Okada K (2024) Predation facilitates evolution of sex-specific antipredator defences in a sexually dimorphic beetle. Anim Behav 215:81–88

Khelifa R (2017a) Faking death to avoid male coercion: extreme sexual conflict resolution in a dragonfly. Ecology 98:1724–1726

Khelifa R (2017b) Spatiotemporal pattern of phenology across geographic gradients in insects. Doctor Thesis, University of Zurick

King BH, Leaich HR (2006) Variation in propensity to exhibit thanatosis in *Nasonia vitripennis* (Hymenoptera: Pteromalidae). J Insect Behav 19:241–249

Kiyotake H, Matsumoto H, Nakayama S, Sakai M, Miyatake T, Ryuda M, Hayakawa Y (2014) Gain of long tonic immobility behavioral trait causes the red flour beetle to reduce anti-stress capacity. J Insect Physiol 60:92–97

Klank J, Protti-Sánchez F, Triana F, Stynoski JL (2023) Thanatosis in the floating rocket frog, *Silverstoneia flotator* from Costa Rica. Herpetol Notes 16:349–351

Konishi K, Matsumura K, Sakuno W, Miyatake T (2020) Death feigning as an adaptive anti-predator behaviour: further evidence for its evolution from artificial selection and natural populations. J Evol Biol 33:1120–1128

Krams I, Kivleniece I, Kuusik A, Krama T, Freeberg TM, Mänd R, Sivacova L, Rantala MJ, Mänd M (2014) High repeatability of antipredator responses and resting metabolic rate in a beetle. J Insect Behav 27:57–66

Kreidl A (1916) Uber hypnose bei Fischen. Pfluger's Arch Gener Physiol 164:441–444

Kudo T, Hasegawa E (2022) Relationship between the duration of death-feigning and temperature in black and red elytral types of a polymorphic ladybug, *Harmonia axyridis*. J Insect Behav 35:11–16

Kuriwada T, Kumano N, Shiromoto K, Haraguchi D (2009) Copulation reduces the duration of death-feigning behaviour in the sweetpotato weevil, *Cylas formicarius*. Anim Behav 78:1145–1151

Kuriwada T, Kumano N, Shiromoto K, Haraguchi D (2010a) The effect of mass-rearing on death-feigning behaviour in the sweet potato weevil (Coleoptera: Brentidae). J Appl Entomol 134:652–658

Kuriwada T, Kumano T, Shiromoto N, Haraguchi D (2010b) Effect of irradiation on death-feigning behavior in the male sweetpotato weevil *Cylas formicarius* (Coleoptera: Blentidae). Fla Entomol 93:39–44

Kuriwada T, Kumano N, Shiromoto K, Haraguchi D (2011) Age-dependent investment in death-feigning behaviour in the sweetpotato weevil *Cylas formicarius*. Physiol Entomol 36:149–154

Larsen TB (1991) The art of feigning death-thanatosis in Euploea (Danainae) and other aposematic butterflies. Entomol Rec 103:263–266

Lawrence SE (1992) Sexual cannibalism in the praying mantid, Mantis religiosa: a field study. Anim Behav 43:569–583

Li H, Wen J (2021a) Behaviour and metabolism during tonic immobility (death-feigning) in *Eucryptorrhynchus scrobiculatus* and *E. brandti* (Coleoptera: Curculionidae). Eur J Entomol 118:322–329

Li H, Wen J (2021b) Factors influencing the duration of death feigning in *Eucryptorrhynchus scrobiculatus* and *E. brandti* (Coleoptera: Curculionidae). J Ethol 40:61–69

Li H, Zhang G, Ji Y, Wen J (2018) Effects of starvation on death-feigning in adult *Eucryptorrhynchus brandti* (Coleoptera: Curculionidae). Ethology 125:645–651

Li H, Wang L, Wen J (2023) Observations on neurophysiological pattern and behavioural traits as death-feigning mechanism in *Eucryptorrhynchus serobiculatus* (Coleoptera: Curculionidae). J Exp Biol 226(20):jeb245864

Li R, Yang C, Lou Y, Lü J, Lyu N (2025) Animal personalities influence male mating behavior and reproductive fitness. Behav Ecol 36(1):arae094

Lira AFA, Almedia FM, Alburquerque CMR (2020) Reaction under the risk of predation: effects of age and sexual plasticity on defensive behavior in scorpion Tityus pusillus (Scorpiones: Buthidae). J Ethol 38:13–19

Lissman HW (1946) The neurological basis of the locomotory rhythm in the spinal dogfish (*Scyallum canicula, Acanthias vulgaris*) I. Reflex Behavior. J Exp Zool 23:143–161

Liu J, Liu K, Tang Y, Wang W, Xu X, Liang J, Xiao Y, Elgar MA (2024) Females adopt sexual catalepsy to facilitate mating. Curr Zool 70:174–181

Luitel SC, Devkova K (2023) First report of death-feigning behavior in a twin-spotted wolfsnake, *Lycodon jara* (Shaw 1802), from Nepal. Reptile & Amphibians 30:e19482

Lundgren EJ, Moeller KT (2017) Anti-predator strategies of, and possible thanatosis in, juvenile collared peccaries (*Pecari tajacu*). Southwest Nat 62:235–237

Maekawa T, Ohara K, Zhang Y, Fukutomi M, Matsumoto S, Matsumura K, Shidara H, Yamazaki S, Fujisawa R, Ide K, Nagaya N, Yamazaki K, Koike S, Miyatake T, Kimura K, Ogawa H, Takahashi S, Yoda K (2020) Deep learning-assisted comparative analysis of animal trajectories with DeepHL. Nat Commun 11:Article number: 5316. https://www.nature.com/articles/s41467-020-19105-0

Maekawa T, Higashide D, Hara T, Matsumura K, Ide K, Miyatake T, Kimura K, Takahashi S (2021) Cross-species behavior analysis with attention-based domain-adversarial Deep Neural Networks. Nat Commun 12:5519. https://doi.org/10.1038/s41467-021-25636-x

Mangold E (1920) Die tierische hypnose (Einschliesslich tonisch tetanishe und tostell-reflaex: reactions-Akunese der Protisen) Ergelon Physiologie 18:79–117

Marques OA, Banci KRS, Strüssmann C (2013) Death-feigning behaviour in water snakes of the genus *Hydrodynastes* (Dipsadidae) from South America. Herpetol Notes 6:95–96

Martinez A, Ritzi CM (2020) Duration of thanatosis is based on temperature in *Estigmene acrea*. Southwest Entomol 45:289–291

Matias CSL, de Araújo DS, de Castro DP (2024) Death simulation behavior of the lizard *Lygodactylus klugei* of the Gekkonidae (Reptilia: Squamata) family in northeastern Brazil. Boletim do Museu Paraense Emílio Goeldi. Ciências Naturais 19:e2024–ee928

Matsumura K (2025) Synchronising anti-predator behaviour in the red flour beetle *Tribolium castaneum*. Behav Ecol 36(2):araf013

Matsumura K, Miyatake T (2015) Differences in attack avoidance and mating success between strains artificially selected for dispersal distance in *Tribolium castaneum*. PLoS One 10(5):e0127042

Matsumura K, Miyatake T (2018) Responses to relaxed and reverse selection in strains artificially selected for duration of death-feigning behavior in the red flour beetle, *Tribolium castaneum*. J Ethol 36:161–168

Matsumura K, Miyatake T (2022) Polygene control and trait dominance in death-feigning syndrome in the red flour beetle *Tribolium castaneum*. Behav Genet 52:298–305

Matsumura K, Miyatake T (2023a) Latitudinal cline of death-feigning behaviour in a beetle (Tribolium castaneum). Biol Lett 19:20230028. https://doi.org/10.1098/rsbl.2023.0028

Matsumura K, Miyatake T (2023b) Dominance and inheritance patterns of mobility and death feigning in beetle strains selected for moving activity. Genetica 15:1–10

Matsumura K, Sasaki K, Miyatake T (2016) Correlated responses in death-feigning behavior, activity, and brain biogenic amine expression in red flour beetle *Tribolium castaneum* strains selected for walking distance. J Ethol 34:97–105

Matsumura K, Yumise K, Fujii Y, Hayashi T, Miyatake T (2020) Anti-predator behaviour depends on male weapon size. Biol Lett 16:20200601. https://doi.org/10.1098/rsbl.2020.0601

Matsumura K, Hosken D, Noda T, Miyatake T, Sharma MD (2025) Behavior, behavioral syndromes, and metabolism: the effects of artificial selection for death-feigning on metabolic rate. J Insect Sci 25:16; ieaf007

McKaye KR (1981) Field observation on death feigning: a unique hunting behavior by the predatory cichlid, *Haplochromis livingstoni*, of Lake Malawi. Environ Biol Fish 6:361–365

Melleu FF, Lino-de-Oliveira C, Marino-Neto J (2017) The mesencephalic GCt–ICo complex and tonic immobility in pigeons (*Columba livia*): a c-Fos study. Brain Struct Funct 222:1253–1265

Messler A, Wund MA, Barker JA, Foster SA (2007) The effect of relaxed and reversed selection by predators on the antipredator behavior of the three stickle back, *Gasteroteus aculeatus*. Ethology 113:953–963

Miura Y, Takii H, Matsumura K, Miyatake T (2025) Experimental evolution of prey by predation pressure: death-feigning duration becomes longer in the presence of predators. Appl Entomol Zool 60:257–263. https://doi.org/10.1007/s13355-025-00913-z

Miyatake T (2001a) Diurnal periodicity of death-feigning in *Cylas formicarius* (Coleoptera: Brentidae). J Insect Behav 14:421–432

Miyatake T (2001b) Effects of starvation on death-feigning in adults of *Cylas formicarius* (Coleoptera: Brenridae). Ann Entomol Soc Am 94:612–616

Miyatake T (2021) Environmental, physiological, and genetic effects on tonic immobility in beetles. In: Sakai M (ed) Death-feigning in insects, entomology monographs. Springer Nature Singapore Pte Ltd, Singapore, pp 39–54

Miyatake T, Katayama K, Takeda Y, Nakashima A, Sugita A, Mizumoto M (2004) Is death-feigning adaptive? Heritable variation in fitness difference of death-feigning behaviour. Proc R Soc Lond Ser B 271:2293–2296

Miyatake T, Okada K, Harano T (2008) Negative relationship between ambient temperature and death-feigning intensity in adult *Callosobruchus maculatus* and *Callosobruchus chinensis*. Physiol Entomol 33:83–88

Miyatake T, Matsumura K, Kitayama R, Otsuki K, Yuhao J, Fujisawa R, Nagaya N (2019) Arousal from tonic immobility by vibration stimulus. Behav Genet 49:478–483

Miyatake T, Abe MS, Matsumura K, Yoshii T (2022) Artificial selections for death-feigning behavior in beetles show correlated responses in amplitude of circadian rhythms, but the period of the rhythm does not. Ethology 128:453–460

Moore AU, Amstey MS (1962) Tonic immobility: differences in susceptibility of experimental and normal sheep and goats. Science 135:729–730

Moore KA, Williams DD (1990) Novel strategies in the complex defense repertoire of a stonefly (*Pteronarcys dorsata*) nymphs. Oikos 57:49–56

Moriya S, Miyatake T (2001) Eradication programs of two sweetpotato pests, *Cylas formicarius* and *Euscepes postfasciatus*, in Japan with special reference to their dispersal ability. JARQ 35:227–234

Moriyama T (2004) Problem solving and autonomous behavior in pill bugs (*Armadillidium vulgare*). Ecol Psychol 16:287–302

Mukharror DA, Susiloningtyas D, Ichsan M (2020) Tonic immobility induction and duration on halmahera walking shark (*Hemischyllium halmahera*). IOP Conf Ser: Earth Environ Sci 404:012080

Muller MMP, Meacham M, Shepard DB, Santana DJ (2024) Defensive displays of speckled Kingsnake, *Lampropeltis holbrooki* Stejneger, 1902. Herpetology 17:217–220

Muscat E, Rotenberg EL, Machado IF (2016) Death-feigning behaviour in an *Erythrolamprus miliaris* (Linnaeus 1758) water snake in Ubatuba, São Paulo, southeastern Brazil (Dipsadidae). Herpetol Notes 9:95–97

Nagaya N, Mizumoto N, Abe MS, Dobata S, Sato R, Fujisawa R (2017) Anomalous diffusion on the servosphere: a potential tool for detecting inherent organismal movement patterns. PLoS One 12(6):e0177480. https://doi.org/10.1371/journal.pone.0177480

Nakamuta K, van Tol WHM, Visser JH (2005) An olfactometer for analyzing olfactory responses of death-feigning insects. Appl Entomol Zool 40:173–175

Nakano R (2021) Ultrasound-induced freezing response in moths. In: Sakai M (ed) Death-feigning in insects, entomology monographs. Springer Nature Singapore Pte Ltd, Singapore, pp 55–72

Nakao K, Sumi T, Matsumura K (2025) Antibiotic treatment eliminated *Wolbachia* in *Tribolium confusum* (Coleoptera: Tenebrionidae) and prolonged tonic immobility duration in later generations. https://doi.org/10.1007/s13355-025-00915-x

Nakayama S, Miyatake T (2009a) Positive genetic correlations between life-history traits and death-feigning behavior in adzuki bean beetle (*Callosobruchus chinensis*). Evol Ecol 23:711–722

Nakayama S, Miyatake T (2009b) Genetic trade-off between abilities to avoid attack and to mate: a cost of tonic immobility. Biol Lett 6:18–20

Nakayama S, Miyatake T (2010) A behavioral syndrome in the adzuki bean beetle: genetic correlation among death feigning, activity, and mating behavior. Ethology 116:108–112

Namiki S (2021) Descending neuron for freezing behavior in *Drosophila melanogaster*. In: Sakai M (ed) Death-feigning in insects, entomology monographs. Springer Nature Singapore Pte Ltd, Singapore, pp 145–157

Narayan EJ, Cockrem JF, Hero J-M (2013) Sight of a predator induces a corticosterone stress response and generates fear in an amphibian. PLoS One 8(8):e73564

Natta G, Laini A, Roggero A, Fabbriciani F, Rolando A, Palestrini C (2023) Behavioural repeatability and behavioural syndrome in the dung beetle *Copris umbilicatus* (Coleoptera, Scarabaeidae). Insects 14:529. https://doi.org/10.3390/insects14060529

Neves FM, Pie MR (2018) On the adult behavioral repertoire of the sawfly *Perreyia flavipes* Konow, 1899 (hymenoptera: Pergidae): movement, mating, and thanatosis. Neotrop Entomol 47:46–52

Nishi Y, Sasaki K, Miyatake T (2010) Biogenic amines, caffeine and tonic immobility in *Tribolium castaneum*. J Insect Physiol 56:622–628

Nishino H (2004) Motor output characterizing thanatosis in the cricket *Gryllus bimaculatus*. J Exp Biol 207:3899–3915

Nishino H (2021) Tonic immobility in a cricket: neuronal underpinnings of global motor inhibition. In: Sakai M (ed) Death-feigning in insects, entomology monographs. Springer Nature Singapore Pte Ltd, Singapore, pp 109–133

Nishino H, Sakai M (1996) Behaviorally signifcant immobility so called of thanatosis in the cricket *Gryllus bimaculatus* DeGeer. J Comp Physiol 179:613–624

Nishino H, Sakai M (2021) Tonic immobility in a cricket: behavioral characteristics, neural substrate, and functional significance. In: Sakai M (ed) Death-feigning in insects, entomology monographs. Springer Nature Singapore Pte Ltd, Singapore, pp 93–108

Nishino H, Sakai M, Field LH (1999) Two antagonistic functions of neural groups of the femoral chordotonal organ underlie thanatosis in the cricket *Gryllus bimaculatus* DeGeer. J Comp Physiol A 185:143–155

O'Brien TJ, Dunlap WP (1975) Tonic immobility in the blue crab (*Callinectes sapidus*, Rathbun) its relation to threat of predation. J Comp Physiol Psychol 89:86–94

O'Donnell RP (2024) Distribution of the Arizona night lizard, *Xantusia arizonae*. Sonoran Herpetol 37:156–159

Ogundeji T, Ayo JO (2025) Open field and tonic immobility responses in broiler chickens administered lycopene. Vet Integr Sci 23:e2025023-1-10

Ohno T, Miyatake T (2007) Drop or fly? Negative genetic correlation between death-feigning intensity and flying ability as alternative anti-predator strategies. Proc R Soc B 274:555–560

Okai K, Sugiura S (2023) Death feigning in adult diving beetles (Coleoptera: Dytiscidae). J Asia Pac Entomol 26(2):102082. https://doi.org/10.1016/j.aspen.2023.102082

Oliver MK (1996) Death-feigning observed in *Hippopsis lemniscata* (Fabricius) (Coleoptera: Cerambycidae). Coleopt Bull 50:160–161

Örstan A, Novák M (2022) A firefly (Lampyridae) larva feeding on a snail. Bull Amateur Entomol Soc 81:184–186

Ouyang H, Wu P, Zhang R, Haseeb M (2020) Trade-off investment between tonic immobility and mate search in the Sweetpotato weevil, *Cylas formicarius* (Coleoptera: Brentidae). Insects 11:774. https://doi.org/10.3390/insects11110774

Ouyang H, Zhang R, Haseeb M (2023) Investment trade-off between mating behavior and tonic immobility in the Sweetpotato weevil *Cylas formicarius* (Coleoptera: Brentidae). Insects 14:73. https://doi.org/10.3390/insects14010073

Oyen KJ, Croucher L, Benoit JB (2021) Tonic immobility is influenced by starvation, life stage, and body mass in ixodid ticks. J Med Entomol 58:1030–1040

Özentürk U (2024) Effect of environmental enrichment within cages on production and welfare in laying hens. Braz J Poultry Sci 26:1–13

Páez AM, Padilla EMH, Klimley AP (2023) A review of tonic immobility as an adaptive behavior in sharks. Environ Biol Fish 106:1455–1462

Pandey S, Samal A, Pattnaik B (2023) Thanatosis behavior: a unique strategy for survival by a spectacled cobra, *Naja naja* (Linnaeus 1758) (Squamata: Elapidae) in Bhubaneswar, Odisha. Entomol Ornithol Herpetol Curr Res 11:1000292

Passos LF, Garcia G, Young RJ (2017) The tonic immobility test: do wild and captive golden mantella frogs (*Mantella aurantiaca*) have the same response? PLoS One 12:e0181972

Patel H, Naik V, Tank SK (2016) Death-feigning behavior in two species of *Lygosoma* (Squamata: Scincidae) from India. Phyllomedusa J Herpetol 15:191

Peraza-Sánchez AJ (2024) Merocoris (Corynocoris) distinctus Dallas, 1852 (Coreidae: Meropachyinae: Merocorini): first record for Costa Rica with a unique case of thanatosis. Entomological Communications 6:ec06013

Pereyra P, Saraco M, Maldonado H (1999) Decreased response or alternative defensive strategies in escape: two different types of long-term memory in the crab *Chasmagnathus*. J Comp Physiol A 184:301–310

Petit S, Hammond PA, Heterick B, Weyland JJ (2023) *Polyrhachis femorata* (hymenoptera: Formicidae) habitat and colony defensive immobility strategy. Aust J Zool 70:126–131

Pflüger H, Büschges A, Bässler U (2021) Historical review on Thanatosis with special reference to the work of Fritz Steiniger. In: Sakai M (ed) Death-feigning in insects, entomology monographs. Springer Nature Singapore Pte Ltd, Singapore, pp 15–21

Prohammer LA, Wade MJ (1981) Geographic and genetic variation in death-feigning behavior in the flour beetle, *Tribolium castaneum*. Behav Genet 11:395–401

Pyle P, Schramm MJ, Keiper C, Anderson SD (1999) Predation on a white shark (Carcharodon carcharias) by a killer whale (Orcinus orca) and a possible case of competitive displacement. Mar Mamm Sci 15:563–568

Quadros AF, Bugs PS, Araujo PB (2012) Tonic immobility in terrestrial isopods: intraspecific and interspecific variability. Zookeys 20:155–170

Radabaugh DC (1989) Seasonal colour changes and shifting antipredator tactics in darters. J Fish Biol 34:679–685

Ramalho WP, Guerra V, Ferraz D, Machado IF, Vieira L (2019) Observations on death-feigning behaviour and colouration patterns as anti-predator mechanisms in Amazonian anurans. Herpetol Notes 12:269–272

Ramos G, Vital VAF, Jardim THA, Nunes G, Branco MEC, de Azevedo CS, Sant'Anna AC (2023) Individual responses of captive amazon parrots to routine handling can reflect their temperament. Animals 13:738. https://doi.org/10.3390/ani13040738

Reitze M, Nentwig W (1991) Comparative investigations into the feeding ecology of six Mantodea species. Oecologia 86:568–574

Resende AC, Streatfield J, Rogers A (2025) Thanatosis in juvenile *notolabrus celidotus*, the New Zealand Spotty Wrasse. acta ethol 28:45–48

Ritter C, de Mol F, Richter E, Struck C, Katroschan K (2016) Antipredator behavioral traits of some *Agriotes* wireworms (Coleoptera: Elateridae) and their potential implications for species identification. J Insect Behav 29:214–232

Rogers SM, Simpson SJ (2014) Thanatosis. Curr Biol 24:R1031–R1033

Rojas-Suárez AJ, Estrada S, Montes-Correa AC, Jiménez-Bolaño JD (2023) Death-feigning in the red-bellied ground lizard, *Stenocercus erythrogaster* (Hallowell 1856) (Squamata: Tropiduridae), in Colombia. Reptiles Amphibians 30:e20083

Rusli MU, Wu NC, Booth DT (2016) Tonic immobility in newly emerged sea turtle hatchlings. Chelonian Conserv Biol 15:143–147

Sabelis NW, van Rijn CJ (2002) Predation by insects and mites. In the impact of supplementary food on a prey - predator interaction, Thrips as crop pests. University of Amsterdam

Sakai M (2021a) Death-feigning in insects:mechanism and function of tonic immobility. Springer Nature Singapore Pte Ltd, Singapore

Sakai M (2021b) Freezing and tonic immobility: their definitions and naming. In: Sakai M (ed) Death-feigning in insects, entomology monographs. Springer Nature Singapore Pte Ltd, Singapore, pp 1–14

Sanchéz Paniagua K, Abarca JG (2016) Thanatosis in four poorly known toads of the genus *Incilius* (Amphibia: Anura) from the highlands of Costa Rica. Mesoamerican Herpetol 3:135–140

Sannolo M, Gatti F, Scali S (2014) First record of thanatosis behaviour in *Malpolon monspessulanus* (Squamata: Colubridae). Herpetol Notes 7:323

Santos MB, Oliveira MCLM, Verrastro L, Tozetti AM (2010) Playing dead to stay alive: death-feigning in *Liolaemus occipitalis* (Squamata: Liolaemidae). Biota Neotropica 10:361–364

Sargeant AB, Eberhardt LE (1975) Death feigning by ducks in response to predation by red foxes (*Vulpes fulva*). Am Midl Nat 94:108–119

Sato DX, Takahashi Y (2025) Neurogenomic and behavioral principles shape freezing dynamics and synergistic performance in *Drosophila melanogaster*. Nat Commun 6:5928. https://doi.org/10.1038/s41467-025-61313-z

Saxena SC (1957) An experimental study of thanatosis in *Armadillidium vulgare* (Latreille). J Zool Soc India 9:192–199

Scarton LP, Zimmermann BL, Machado S, Aued AW, Manfio D, Santos S (2009) Thanatosis in the freshwater crab *Trichodactylus panoplus* (Decapoda: Brachyura: Trichodactylidae). Nauplius 17:87–100

Schaefer JG (1921) Uber den lagerreflextonus von *Raja clavata*. Biol Zentrabl 41:291–295

Schulte L, Caspers BA (2024) Tonic immobility behaviour does not differ between fire salamander larvae from ponds and streams. Ecol Evol 14:e11211

Segovia JMG, Moura RR, Willemart RH (2019a) Starvation decreases behavioral consistency in a Neotropical harvestman. acta ethologica 22(203):208

Segovia JMG, Murayama GP, Willemari RH (2019b) Sexual differences in weaponry and defensive behavior in a neotropical harvestman. Curr Zool 65:553–558

Sendova-Franks AB, Worley A, Franks NR (2020) Post-contact immobility and half-lives that save lives. Proc R Soc Lond B Biol Sci 287:20200881. https://doi.org/10.1098/rspb.2020.0881

Shreeve TG, Dennis RLH, Wakeham-Dawson A (2006) Phylogenetic, habitat, and behavioural aspects of possum behaviour in European Lepidoptera. J Res Lepid 39:80–85

Simmons LW (2001) Sperm competition and its evolutionary consequences in the insects. Princeton University Press, p 448

Skok J, Povše MP, Škorjanc D (2025) A case of thanatosis in domestic sheep. Behaviour. https://doi.org/10.1163/1568539X-bja10311

Skomorucha I, Sosnówka-Czajka E (2025) The effects of origin on the formation of stress parameters and selected behaviour of hens. Agriculture 15:95

Spencer CP, Richard K (2013) Are invertebrate pedestrians threatened? Observations of *Hoplogonus simsoni* from road line transects in northeastern Tasmania. Tasmanian Nat 135:28–40

Suzuki S, Futami K (2017) Predatory risk increased due to egg-brooding in *Armadillidium vulgare* (isopoda: Oniscidea). Ethology 124:256–259

Suzuki K, Ikebuchi M, Okanoya K (2013) The impact of domestication on fearfulness: a comparison of tonic immobility reactions in wild and domesticated finches. Behav Process 100:58–63

Takanashi T, Kojima W (2021) Vibration-induced immobility in coleopteran insects. In: Sakai M (ed) Death-feigning in insects, entomology monographs. Springer Nature Singapore Pte Ltd, Singapore, pp 73–92

Tanaka K, Sasaki S, Matsumura K, Yajima S, Miyatake T (2021) Genomic characterization between strains selected for death-feigning duration for avoiding attack of a beetle. Sci Rep 11:1–18. https://doi.org/10.1038/s41598-021-00987-z

Taylor KL, Henry CS, Farkas TE (2023) Why fake death? Environmental and genetic control of tonic immobility in larval lacewings (Neuroptera: Chrysopidae). J Insect Sci 23:1–5

Templeton CN, Shriner WM (2004) Multiple selection pressures influence Trinidadian guppy (*Poecilia reticulata*) antipredator behavior. Behav Ecol 15:673–678

Thompson RKR, Foltin RW, Sweet BA, Graves CA, Lowitz CE (1981) Tonic immobility in Japanese quail can reduce the probability of sustained attack by cats. Anim Lean Behav 9:145–149

Tobler M (2005) Feigning death in the central American cichlid *Parachromis friedrichsthalii*. J Fish Biol 66:877–881

Tojo S (1991) Variation in phase polymorphism in the common cutworm, *Spodoptera litura* (Lepidoptera:Noctuidae). Appl Entomol Zool 26:571–178

Tojo S, Morita M, Himura K (1985) Effects of juvenile hormone on some phase characteristics in the cotton cutworm, *Spodoptera litura*. J Insect Physiol 31:243–249

Toledo LF, Sazima I, Haddad CFB (2010) Is it all death feigning? Case in anurans. J Nat Hist 44:1979–1988

Toledo LF, Sazima I, Haddad CFB (2011) Behavioural defences of anurans: an overview. Ethol Ecol Evol 23:1–25

Triana F, Bonilla F, Alfaro-Chinchilla A, Víquez C, Díaz C, Sasa M (2022) Report of thanatosis in the central American scorpions *Tityus ocelote* and *Ananteris platnicki* (Scorpiones: Buthidae). Euscorpius 359:1–5

Tuohy E, Schärer-Umpierre M, Penrod L, Appeldoorn R (2023) Spatial and temporal dynamics of a Nassau grouper fish spawning aggregation located on an isolated seamount in Puerto Rico. Aquat Conserv Mar Freshwat Ecosyst 33:1116–1130

Uchiyama H, Sasaki K, Hinosawa S, Tanaka K, Matsumura K, Yajima S, Miyatake T (2019) Transcriptomic comparison between beetle strains selected for short and long durations of death feigning. Sci Rep 9:14001. https://doi.org/10.1038/s41598-019-50440-5

Ulate Gómez TN, Welch AE, Peters WS (2024) Short communication: problems of classifying predator-induced prey immobility—an unexpected case of post-contact freezing. Web Ecol 24:35–40. https://doi.org/10.5194/we-24-35-2024

Van Poucke E, Suchánková H, Jensen P (2023) Commercial hatchery processing may affect susceptibility to stress in laying hens. PLoS One 18:e0291324

Van Poucke E, Mitschler P, Jensen P (2024) Play stimulation as a possible means to buffer early stress from commercial hatchery treatment in laying hen chicks. Appl Anim Behav Sci 274:106264

van Veen JW, Sommeijer MJ, Monge IA (1999) Behavioural development and abdomen inflation of gynes and newly mated queens of *Melipona beecheii* (Apidae, Meliponinae). Insect Soc 46:361–365

Vega-Román E, Collado GA, Pizarro-Araya J (2024) New report of thanatosis in a south American myriapod. Entomol Am 130:28–32

Velasco IG, de Oliveira JVA, da Silva Maia NVN, Silva SN, de Oliveira Braga JF, Simões PI (2025) Limbs interweaving for Neotropical anurans. Cuad Herpetol 39(1):00–00

Villet MH (1999) The cicada genus *Nyara n.* gen. (Homoptera: Cicadidae): Systematics, behavior and conservation status. Trop Zool 12:157–163

Vogel HH (1950) Observations on social behavior in *Turkey vultures*. Auk 67:210–216

Volchan E, Souza GG, Franklin CM, Norte C, Rocha-Rego V, Oliveira JM, David IA, Mendlowicz MV, Coutinho ESF, Fiszman A, Berger W, Marques-Portella C, Figueira I (2011) Is there tonic immobility in humans? Biological evidence from victims of traumatic stress. Biol Psychol 88:13–19

Wang L, Li H, Wen J (2023) Instantaneous characteristics of the death-feigning and arousal behavior of *Eucryptorrhynchus Scrobiculatus* and *E. Brandti* adults. J Insect Behav 36:239–246

Watsky MA, Gruber SH (1990) Induction and duration of tonic immobility in the lemon shark *Negaprion brevirostris*. Fish Physiol Biochem 8:207–210

Whishaw IQ, Previsich N, Flannigan KP (1978) Tonic immobility in feral and domestic Dutch rabbits (*Oryctolagus cuniculus*), mountain cottontail (*Sylvilagus nuttalli*), and whitetail jackrabbit (*Lepus townsendi*) as a function of posture. Behav Biol 24:88–96

Whitman PA, Marshal JA, Keller BC Jr (1986) Tonic immobility in the smooth dogfish shark, *Mustelus canis* (Pisces, Carcharinidae). Copeia 3:829–832

Wildermuth H, Jödicke R, Shröter A, Borkenstein A (2019) Death feigning in sexual conflict between dragonflies (Odonata): does it exist? Odonatologica 48:211–228

Williamson MJ, Dudgeon C, Slade R (2018) Tonic immobility in the zebra shark, *Stegostoma fasciatum*, and its use for capture methodology. Env Biol Fish 101:741–748

Yamada S, Urabe J (2021) Death-feigning behaviours increase survival rate of littoral cladocerans under predation by odonate larvae. Freshw Biol 66:2030–2037

Yoder JA, Benoit JB, Bundy MR, Hedges BZ, Gribbins KM (2009) Functional morphology of secretion by the large wax glands (Sensilla sagittiformia) involved in tick defense. Psyche 2009:1–8

Yoshida M (2021) Immobility behaviors in fish: a comparison with other vertebrates. In: Sakai M (ed) Death-feigning in insects, entomology monographs. Springer Nature Singapore Pte Ltd, Singapore, pp 159–178

Factors Affecting Death Feigning

4

Abstract

Fabre, the author of Souvenirs Entomologiques, was troubled by the question of what determines the duration of death feigning. I sometimes wonder what he might have discovered if he had access to modern technology. Do they play dead or not (frequency)? Or how long they continue to feign death (duration) varies enormously, not only between species but also within species. Even if an animal feigns death, it cannot continue doing so indefinitely. If it did, it would likely dry out under the scorching midday sun. More importantly, it could end up being eaten by a predator. For example, ants might carry a small insect feigning death back to their nest. There is a limit to how long death feigning can be maintained—and perhaps there is even an optimal duration for it. In this chapter, I provide an overview of 23 factors that influence the frequency and duration of death-feigning behavior. Among these factors are environmental and physiological ones, such as the time of day during which death feigning occurs—which puzzled Fabre—and temperature. If devices like incubators and precision balances had existed in his time, Fabre would no doubt have conducted all the experiments described in this chapter. I would like to dedicate this chapter to him.

Keywords

Death-feigning duration · Two-modes · Stimulus: temperature · Environmental factors: arousal · Pheromone

4.1 Phenotypic Plasticity: Two Modes in Insects

Previous studies, primarily conducted on beetle species, suggest that insects may generally exhibit two distinct behavioral states: an "active" state and a "non-active" state. Based on this, we can propose the following hypothesis: during the active

© The Author(s), under exclusive license to Springer Nature Singapore Pte Ltd. 2026

T. Miyatake, *Death Feigning*, Ecological Research Monographs,
https://doi.org/10.1007/978-981-95-5319-8_4

state, defensive behaviors such as running or flying serve as effective strategies against predators, whereas in the non-active state, immobility—such as tonic immobility or freezing (see Chap. 2)—functions as a more adaptive antipredator response.

This hypothesis (originally proposed by Miyatake 2001a) was developed based on observations of insects exhibiting different physiological and environmental conditions. These include (1) pre-stimulus behaviors such as walking, resting, or feeding; (2) circadian rhythm; (3) mating status; (4) starvation; (5) ambient temperature; (6) body size; and (7) age, among others. These factors collectively suggest that tonic immobility exhibits phenotypic plasticity, as discussed below. In the following sections, I will examine the influence of these factors, using specific insect examples to illustrate their relevance.

4.1.1 Pre-Stimulus Behaviors

Pre-stimulus behavior significantly influences both the likelihood and duration of tonic immobility in the sweet potato weevil, *Cylas formicarius* (Miyatake 2001a). In this nocturnal species, individuals that were resting prior to stimulation exhibited a higher incidence and longer duration of tonic immobility compared to those that were walking or feeding. These findings support the hypothesis that insects possess two behavioral modes—active and inactive—which are closely associated with the expression of tonic immobility. Specifically, individuals in an inactive or resting state are more prone to exhibit tonic immobility, whereas those in an active state are less likely to do so. The behavior of *C. formicarius* thus provides empirical support for this dual-mode hypothesis. One possible explanation is that active individuals, having already surveyed their surroundings, may be more likely to flee toward known refuges, while resting individuals may adopt tonic immobility as a more efficient immediate defense strategy (Miyatake 2001a) (Fig. 4.1).

Tojo (1991) reported that larvae of the tobacco cutworm, *Spodoptera litura*, reared under high-density conditions exhibited increased melanization and a reduced frequency of death-feigning behavior. This reduction in tonic immobility is likely attributable to frequent tactile stimulation resulting from body contact among individuals in the gregarious phase, which may suppress the expression of death feigning.

4.1.2 Diurnal and Circadian Rhythm

Numerous insect behaviors are regulated by endogenous circadian rhythms governed by the biological clock (Saunders 2002), and tonic immobility is no exception. A classical study by Hoagland (1927) demonstrated a circadian pattern in curling and immobility behaviors in the pill bug *Cylisticus convexus*. Thanatosis has also been reported in two freshwater crustaceans, *Aegla longirostri* and *Aegla denticulata denticulata* (Aeglidae: Tanunculidae), although no evidence of photoperiodic regulation was observed in these species (Coutinho et al. 2013). In contrast, the

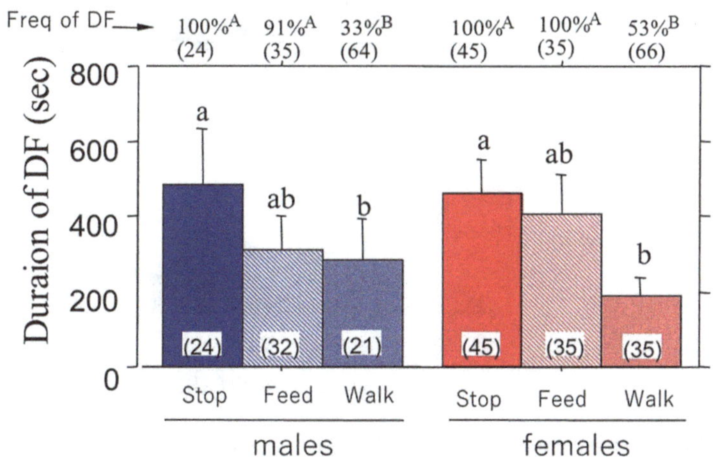

Fig. 4.1 Effects of prestimulus behavior on the duration of death-feint. Filled, hatched, and open bars show durations of death-feint for resting, feeding, and walking weevils, respectively. Numerals in parentheses show the number of weevils that feigned death. Durations under the same alphabetic letter do not differ significantly from each other at the 5% level, as determined by the Scheffe method

scorpion *Tityus pusillus* exhibits a diel variation in thanatosis, with significantly longer durations occurring during the daytime compared to nighttime (Lira et al. 2020).

The expression of tonic immobility behavior is influenced by the time of day. In the sweet potato weevil, *Cylas formicarius*, pre-stimulus behavior significantly affects both the likelihood and duration of tonic immobility (Miyatake 2001a). The frequency of tonic immobility was lower during the daytime and higher at night. Interestingly, even among individuals that exhibited tonic immobility at night, the duration of immobility was shorter compared to those tested during the day (Miyatake 2001a). These findings underscore the importance of considering time-of-day effects—particularly in species with circadian rhythms—when designing and interpreting experiments on tonic immobility (Fig. 4.2).

Why do male weevils exhibit reduced tonic immobility during the nighttime? Two main explanations have been proposed: (1) Predation pressure differs between day and night, potentially altering the adaptive value of immobility; and (2) because *Cylas formicarius* is a nocturnal species, males are primarily engaged in mate searching during the night. Thus, expressing tonic immobility during this period may incur a fitness cost by reducing mating opportunities. A similar pattern has been reported in the seed beetle *Zabrotes subfasciatus*, in which tonic immobility is more frequently observed during the daytime than at night (Cardoso and Mendonça 2019).

The relationship between circadian rhythms and tonic immobility remains an important topic for future investigation, particularly given that tonic immobility is closely associated with locomotor activity (Miyatake 2001a). In other weevil species, such as *Eucryptorrhynchus scrobiculatus* and *E. brandti*, the duration of

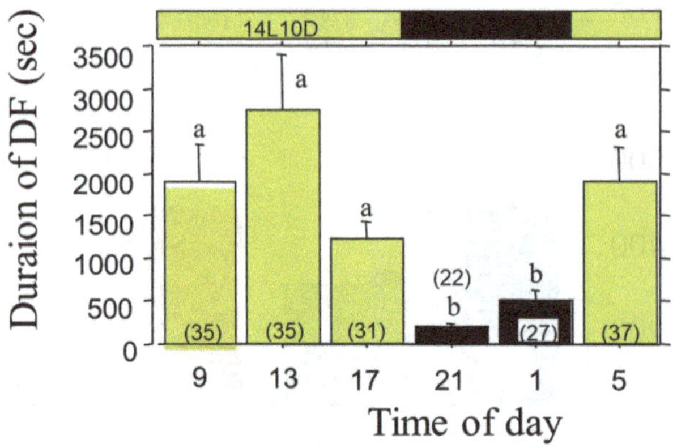

Fig. 4.2 Effects of the diurnal period on the duration of death feigning in Replicate 1. Males and females were assayed for the behavior at the same time in a laboratory. Mean and SE durations of death feigning are shown. Numerals in parentheses show the number of weevils that feigned death. Durations under the same alphabetic letter do not differ significantly from each other at the 5% level as determined by the Scheffe method

thanatosis has been reported to be shorter during the daytime than at night (Li and Wen 2021), suggesting species-specific patterns in the temporal regulation of this behavior.

4.1.3 Mating

In *Cylas formicarius*, copulation has been shown to reduce the duration of tonic immobility in both males and females (Kuriwada et al. 2009). Conversely, insemination was found to increase the duration of tonic immobility in this species (Kuriwada et al. 2009). The effects of mating history on tonic immobility have also been investigated. Notably, males that had previously copulated did not exhibit a significant change in tonic immobility duration (Kuriwada et al. 2009).

However, inseminated females significantly reduced the duration of tonic immobility compared to virgins (Kuriwada et al. 2009). Conversely, mated females exhibited longer durations of tonic immobility than virgin females in *Cylas formicarius* (Kuriwada et al. 2011). Ouyang et al. (2020) investigated the trade-off between thanatosis and mate searching in *C. formicarius* and found that males engaged in mate searching displayed significantly shorter durations of thanatosis than non-searching males, suggesting a negative correlation—or trade-off—between mate searching and thanatosis expression in males. Additionally, the duration of thanatosis was notably reduced during courtship and mating behaviors in this species (Ouyang et al. 2023). To further explore this, Ouyang et al. (2023) performed divergent artificial selection to establish strains with long and short thanatosis durations. Despite clear differences in thanatosis duration between these strains, no significant differences were observed in courtship frequency or reproductive success.

Precopulatory sexual cannibalism by females has been shown to influence male anti-partner behaviors (Bilde et al. 2006). However, the relationship between mating experience and tonic immobility remains understudied and is largely restricted to specific taxa or behavioral contexts, indicating a need for further research.

Dittrich and Rödel (2023) reported that female common frogs (*Rana temporaria*) employ multiple avoidance strategies—namely, "rotation," "release calls," and tonic immobility (death feigning)—to evade persistent male courtship and forced mating attempts. Based on their observations, Dittrich and Rödel (2023) suggest that death feigning in this context functions as an active immobilization strategy rather than the previously assumed passive avoidance behavior.

4.1.4 Starvation

Acheampong and Mitchell (1997) found that well-fed Colorado potato beetles (*Leptinotarsa decemlineata*) required a longer time to recover from immobility compared to individuals starved for 24 hours. However, the authors characterized this immobility as quiescence rather than thanatosis, and thus the observed behavior is more accurately interpreted as freezing.

Similarly, in *Cylas formicarius*, starved individuals of both sexes exhibited a higher frequency of thanatosis but shorter durations of immobility compared to well-fed beetles (Miyatake 2001b). Notably, most starved *C. formicarius* individuals were unable to exhibit tonic immobility and displayed significantly shorter thanatosis durations than their well-nourished counterparts (Miyatake 2001b).

Oyen et al. (2021) investigated thanatosis duration in two tick species and found that prolonged starvation led to a decrease in thanatosis expression, accompanied by an increase in aggressive behaviors. In contrast, a study on the harvestman *Mischonyx cuspidatus* reported no significant difference in thanatosis duration between starved and fed individuals (Segovia et al. 2019a).

Taylor et al. (2023) reported that starvation of Chrysopidae larvae for more than 1 day led to an increased frequency of thanatosis accompanied by a reduced duration of the behavior.

Thanatosis behavior has been studied in the snout weevil *Eucryptorrhynchus brandti*, a forest pest in China that also infests *Ailanthus altissima*. This species exhibits immobility by contracting its legs and pressing its body against the substrate (Li et al. 2018). In *E. brandti*, thanatosis intensity—defined by both frequency and duration—was found to be negatively correlated with the duration of starvation; that is, prolonged starvation in adults resulted in shorter thanatosis durations (Li et al. 2018). Li et al. (2018) further reported that this negative relationship between starvation and thanatosis duration was similar in both males and females, consistent with findings in the sweet potato weevil, *Cylas formicarius*, where no sex differences in starvation tolerance were observed. It is important to note that thanatosis intensity encompasses both the likelihood of exhibiting thanatosis and the duration for which it is maintained.

4.1.5 Ambient Temperature

Saxena (1957) observed that woodlice (*Armadillidium vulgare*) exhibited stronger tonic immobility at lower temperatures. The effect of temperature on the intensity of tonic immobility has also been investigated in two coleopteran species, *Callosobruchus maculatus* and *C. chinensis*. Both species demonstrated negative correlations between temperature and both the frequency of individuals exhibiting death-feigning behavior and the duration of tonic immobility (Miyatake et al. 2008). These findings indicate that lower temperatures are associated with prolonged durations of death-feigning behavior.

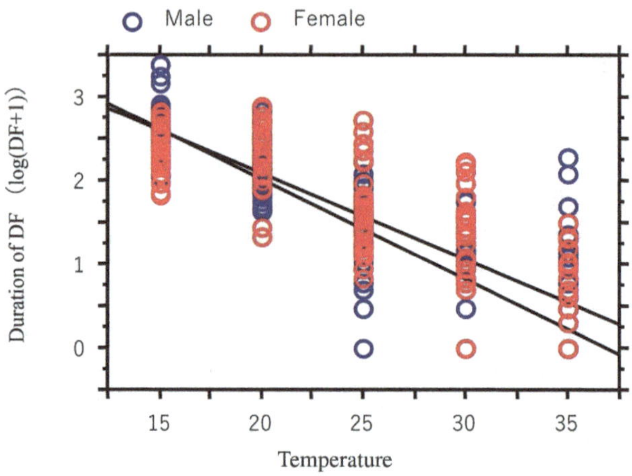

In contrast, tonic immobility occurs more frequently at higher temperatures than at lower temperatures in the seed beetle *Zabrotes subfasciatus* (Cardoso and Mendonça 2019). Similarly, a negative phenotypic correlation between ambient temperature and duration of tonic immobility has been reported in aquatic hemipterans; for example, the water-scorpion *Ranatra* sp. exhibits shorter durations of death-feigning behavior at higher temperatures (Holmes 1906).

Holmes (1906) investigated the effect of temperature on thanatosis duration in the water mantis, comparing high (30 °C) and low (10–14 °C) temperatures, and found that higher temperatures were associated with shorter thanatosis durations. In the parasitic bee *Heterospilus prosopidis*, thanatosis occurs more frequently at lower temperatures (18 °C) than at higher temperatures (25 °C) (Amemiya and Sasakawa 2021).

Kudo and Hasegawa (2022) examined thanatosis duration across three temperatures (25 °C, 20 °C, and 15 °C) in the ladybird *Harmonia axyridis*, which exhibits adult color polymorphism. Their results revealed that in the black (B) morph, thanatosis duration decreased at lower temperatures, whereas the red (R) morph showed no significant temperature-dependent variation, suggesting a complex interaction between temperature and color polymorphism influencing thanatosis behavior.

Li and Wen (2021) studied the relationship between temperature and thanatosis duration in two weevil species, *Eucryptorrhynchus scrobiculatus* and *E. brandti*. They found that in *E. scrobiculatus*, thanatosis duration was reduced at both low (15 °C) and high (32 °C) temperatures, whereas in *E. brandti*, the duration was prolonged at the low temperature (15 °C).

As described above, the relationship between temperature and tonic immobility behavior appears to vary among different taxa, highlighting the need for further investigation. For example, in the grass snake (*Natrix natrix*), no significant correlation has been found between body temperature and the duration of death-feigning behavior (Gregory et al. 2007).

4.1.6 Body Size

In the stick insect *Timema cristinae*, smaller individuals are more likely to exhibit death-feigning behavior than larger individuals; however, this size effect is confounded by the age of wild-caught specimens (Farkas 2016). In the adzuki bean beetle, larger individuals display longer durations of death feigning compared to smaller ones (Hozumi and Miyatake 2005). Among marine isopods, *Cleantiella isopus* males show a negative correlation between body size and duration of death feigning, whereas this relationship is not significant in females (Igarashi and Wada 2024). In males of the horned beetle *Gnatocerus cornutus*, which possess enlarged mandibles used in male-male competition, larger males with larger mandibles are less likely to engage in death-feigning behavior upon stimulation; however, thanatosis duration is not correlated with male size (Matsumura et al. 2020). Despite these findings, studies on isopods remain limited, and further research is needed.

Juvenile European grass snakes (*Natrix natrix*) are less likely to exhibit death-feigning behavior and tend to flee more rapidly in response to human-grabbing stimuli compared to larger conspecifics (Gregory 2008). In contrast, within the same family, dice snakes (*Natrix tessellata*) show an opposite pattern, where larger individuals display longer durations of death feigning than smaller ones (Bjelica et al. 2023).

Farkas (2016) studied death-feigning behavior in the stick insect *Timema cristinae* and found that smaller individuals were more likely to exhibit thanatosis than larger individuals. In males of the horned beetle *Gnatocerus cornutus*, a species in which only males possess enlarged mandibles used for combat, larger males with bigger mandibles were less likely to display thanatosis in response to stimuli; however, thanatosis duration was not correlated with male body size (Matsumura et al. 2020).

Tonic immobility responses in finches have been shown to be unaffected by body weight and growth conditions (Suzuki et al. 2013). More recently, juvenile larval fire salamanders (*Salamandra salamandra*) have been reported to exhibit death-feigning behavior as an antipredator strategy; however, no relationship was found between body size and the duration of death feigning in this species (Schulte and Caspers 2024). Given the variability in findings across taxa and experimental conditions, further research is needed to clarify the relationship between body size and death-feigning behavior.

4.1.7 Sex

Numerous studies have investigated sex-related differences in death-feigning behavior. Three notable studies are summarized as follows. First, Quadros et al. (2012) reported no significant difference between sexes in the duration of immobility in the wharf roach (Ligiidae). Second, Suzuki and Futami (2017) demonstrated that adult female Armadillidium vulgare are unable to exhibit death-feigning behavior while carrying eggs in their abdomen, indicating a sex-specific reproductive cost during incubation. Third, in the marine isopod *Cleantiella isopus*, Igarashi and Wada (2024) observed sex differences in both the frequency and duration of death-feigning behavior; males exhibited greater behavioral variation than females, suggesting that distinct factors influence death-feigning behavior across sexes.

Male mantids of the species Mantis religiosa exhibit freezing behavior after mating to avoid postcopulatory cannibalism by females (Lawrence 1992). Similarly, female robber flies of the species *Efferia varipes* display comparable immobility to reduce harassment by males (Dennis and Lavigne 1976).

In insects, a sex difference in thanatosis has been reported in the parasitoid wasp *Heterospilus prosopidis*, with females exhibiting thanatosis more frequently than males (Amemiya and Sasakawa 2021). However, sex differences in thanatosis behavior in this species vary among local populations: females display thanatosis more often than males in populations derived from Arizona, whereas in populations derived from Hawaii, males exhibit thanatosis more frequently than females (Fujioka et al. 2024). Additionally, sex differences in death-feigning behavior were investigated in the harvestman *Mischonyx cuspidatus*, revealing no significant differences between females and males in either the frequency or duration of thanatosis (Segovia et al. 2019b).

Kudo and Hasegawa (2022) reported no significant sex difference in the duration of death feigning in *Harmonia axyridis*, a ladybird species exhibiting adult color polymorphism. Experiments using strains of the horned beetle *Gnatocerus cornutus*, which differ in sex-specific antipredator behaviors, demonstrated that males predominantly avoid predator attacks through movement, whereas females more frequently employ death-feigning behavior (Katsuki et al. 2024). In contrast, no sex difference in the duration of thanatosis was observed in the adzuki bean weevil *Callosobruchus chinensis* (Hozumi and Miyatake 2005). Furthermore, Odén et al. (2005) showed that social environment influences tonic immobility expression, reporting that female laying hens exhibit shorter durations of tonic immobility than males, potentially due to males acting as guards and thereby reducing fear levels in females.

In the crustacean species *Aegla longirostri* and *Aegla denticulata denticulata* (family Aeglidae, formerly Tanneriidae), females exhibit longer durations of thanatosis compared to males (Coutinho et al. 2013). In contrast, tonic immobility responses in finch species were found to be unaffected by sex, body weight, or growth conditions (Suzuki et al. 2013).

As described above, sex differences in thanatosis vary widely among species and populations, with no consistent pattern emerging to date. Therefore, further research is warranted to elucidate these variations.

4.1.8 Age

Younger ants exhibited thanatosis more frequently than older individuals (Cassill et al. 2008). Kuriwada et al. (2009, 2011) reported that in *Cylas formicarius*, the duration of tonic immobility decreases with age in females but remains unchanged in males. In contrast, no age-related effects on tonic immobility duration were observed in the Colorado potato beetle (Leptinotarsa decemlineata) (Achampong and Mitchell 1997). Additionally, a significant interaction between age and mating history was identified in *C. formicarius* (Kuriwada et al. 2011). Similarly, in the stick insect *Timema cristinae*, younger individuals exhibited thanatosis more frequently than older conspecifics (Farkas 2016). Future studies should examine in detail the relationships among starvation, reproductive age, and mating history in relation to tonic immobility behavior.

In the parasitoid wasp *Heterospilus prosopidis*, thanatosis behavior occurs more frequently at emergence (0 day old) than at 18 days of age, indicating a decline with age (Amemiya and Sasakawa 2021). Recently, Fujioka et al. (2024) demonstrated that the frequency of thanatosis in this species varies with the host bee's daily age across geographic strains, with higher thanatosis frequency observed when the host is younger, and a decrease as the host ages. This pattern may be explained by the reduced flight ability of newly emerged (day 0) and 1-year-old individuals, making thanatosis the most adaptive antipredator strategy during these stages. However, as the natural predators of these parasitoid wasps remain speculative and the experiments were conducted under human-induced stimuli, further field studies are required to elucidate the relationship between predation pressure and thanatosis behavior. Additionally, Fujioka et al. (2024) reported that thanatosis frequency varies among geographic strains depending on the age of the parasitic host, the adzuki bean weevil (*Callosobruchus chinensis*).

Cassill et al. (2008) observed that in colonies of the fire ant *Solenopsis invicta*, younger workers exhibit immobility to avoid conflicts with competing ants. This tendency decreases with age, as older ants are more likely to engage in aggressive interactions rather than remain immobile in the presence of conspecifics. Experimental evidence suggests that immobility enhances survival in young ants, which possess weaker exoskeletons, increasing their survival rate by a factor of four or more (Cassill et al. 2008). In scorpion flies (*Panorpa japonica*), whose larvae display thanatosis behavior, Ishihara et al. (2024) reported that the immobile posture changes ontogenetically: younger larvae adopt a stick-like immobile posture, while older larvae in the third and fourth instars exhibit a ball-like immobility. Similarly, Farkas (2016) demonstrated that younger individuals of the stick insect

Timema cristinae exhibit thanatosis more frequently than older individuals. In other weevil species, *Eucryptorrhynchus scrobiculatus* and *E. brandti*, the duration of thanatosis also decreases with age (Li and Wen 2021).

Similarly, in dice snakes, immature individuals are less likely to exhibit death-feigning behavior compared to adults (Bjelica et al. 2024). Francq (1969) reported that all six examined opossums exhibited death-feigning behavior lasting at least 12 days, with one individual maintaining the behavior for up to 60 days. Additionally, younger individuals were found to be more prone to feigning death.

4.1.9 Season

In *Lycorma delicatula*, antipredator tactics vary seasonally; individuals are more likely to jump away during the early season, whereas death-feigning behavior becomes more frequent in the late season (Kang et al. 2017). This variation is thought to reflect differences in predator communities across seasons. Similarly, in a species of the spider crab genus *Chasmagnathus*, thanatosis occurs more frequently during winter days (Pereyra et al. 1999). Adults of the spotted lanternfly, *Lycorma delicatula*, have been observed to escape by jumping when stimulated early in the outbreak season, but they exhibit significantly increased death-feigning behavior in the late outbreak season (Kang et al. 2017). Despite these observations, the relationship between seasonal changes and death-feigning behavior remains poorly studied, representing an open avenue for research on thanatosis within ecological contexts, including seasonal dynamics.

4.1.10 Morph

In the marine isopod *Cleantiella isopus*, which exhibits body color polymorphism, the duration of immobility was measured across morphs; however, no significant differences were found among morphs (Igarashi and Wada 2024). In dice snakes, individual variation in body size exists, and blotched (spotted) individuals exhibited longer durations of tonic immobility compared to green or melanistic black individuals (Bjelica et al. 2023). In the scorpion species *Tityus pusillus*, the duration of thanatosis has been reported to be longer in adults than in juveniles (Lira et al. 2020).

4.1.11 Background and Ecological Factors

In the parasitoid wasp *Nasonia vitripennis*, variation in background color has been shown to influence both the frequency and duration of death-feigning behavior (King and Leaich 2006). Conversely, in another parasitoid wasp species, *Heterospilus prosopidis*, no effect of background color on the frequency of death feigning has been reported (Amemiya and Sasakawa 2021).

The effect of different substrates (e.g., paper, shallow sand, or deep sand) on the defensive responses of insects—such as tonic immobility, fleeing, or diving for cover—has been investigated in the larvae of the antlion *Euroleon nostras* (Farkas 2016). This study demonstrated that escape behavior varies dynamically depending on the substrate type.

O'Brien and Dunlap (1975) reported that blue crabs (*Callinectes sapidus*) immobilized on sandy substrates, which allow for burial and concealment, exhibited shorter durations of tonic immobility (TI) compared to individuals immobilized on hard surfaces. Holmes (1906) observed that in the water mantis, the duration of death-feigning behavior was shorter under illuminated conditions than in darkness. Similarly, Hennig et al. (1976) found that Carolina anole lizards (*Anolis carolinensis*) exhibited shorter durations of death feigning when immobilized near large bushes compared to those immobilized in open areas lacking protective cover.

4.1.12 Genetic Background

Numerous studies have reported variation in the frequency and duration of death-feigning behavior among local strains, whether collected from the wild or reared in the laboratory. In the parasitic wasp *Heterospilus prosopidis*, differences in the frequency of immobility (pseudoparasitism) were observed among local strains, with Hawaii-reared strains exhibiting a lower frequency compared to those collected in Arizona (Fujioka et al. 2024). Moreover, sib analyses have revealed significant genetic variance in death-feigning frequencies, with notable effects attributable to both sire and dam (Fujioka et al. 2024). Taylor et al. (2023) estimated the broad-sense heritability of the propensity to enter tonic immobility as 0.502 in larvae of the family Chrysopidae.

Farji-Berner et al. (2024b), using larvae of the antlion species *Myrmeleon crudelis*, found significant differences in death-feigning duration between laboratory-reared and field-collected strains, with progressively reared larvae exhibiting significantly shorter durations compared to their wild counterparts. Kudo and Hasegawa (2022) reported that in *Harmonia axyridis*, a ladybird species exhibiting adult color polymorphism, the frequency of death-feigning behavior varied among color morphs; specifically, the red morph displayed longer durations of death feigning than the black morph. Moreover, this difference was modulated by temperature, indicating a temperature-dependent effect on death-feigning behavior.

Adult *Cylas formicarius* weevils that have been mass-reared for 71 generations under sterile insect technique (SIT) programs showed no significant difference in the duration of death-feigning behavior compared to wild adult individuals (Kuriwada et al. 2010a). This finding suggests that death feigning may incur no direct fitness costs and may not evolve without associated trade-offs (although contrasting results have been reported in *Tribolium castaneum*; Nakayama and Miyatake 2009b; Kiyotake et al. 2014). Additionally, comparisons of tonic immobility duration among different chicken strains revealed no significant differences between males and females (Akyol et al. 2024).

Conversely, some studies have indicated that maternal effects, rather than genetic factors, influence death-feigning behavior. Moore and Amstey (1962) experimentally demonstrated that lambs can exhibit tonic immobility, whereas those transferred to foster care at 1 year of age do not, suggesting a significant role of maternal influences. In *Tribolium castaneum*, which will be discussed in a later chapter, extensive research has been conducted on the genetic basis underlying the duration of death-feigning behavior.

Suzuki et al. (2013) measured tonic immobility responses in Bengalese finches subjected to physical restraint to assess fearfulness related to predation coping strategies. Their results demonstrated that Bengalese finches exhibited reduced tonic immobility compared to their ancestor species, the white-backed munia. Additionally, tonic immobility responses were not influenced by sex, body weight, or growth conditions within the finch species (Suzuki et al. 2013). These findings suggest that tonic immobility may be altered through domestication and associated selective pressures over generations. Supporting this, Mignon-Grasteau et al. (2017) reported moderate heritability for tonic immobility traits—such as duration and number of inductions—in broiler chickens (*Gallus gallus domesticus*).

4.1.13 Population Density

In antlion larvae, Farji-Berner et al. (2024b) found that higher population densities—indicated by shorter distances between larval nests in the sandpit—were associated with shorter durations of death feigning in captured larvae. In dice snakes (*Natrix tessellata*), individuals inhabiting islands with low mammalian predator presence tend to exhibit death-feigning behavior rather than slower escape speeds. Moreover, the intensity of death feigning appears to be influenced by the population densities of both heterospecifics and conspecifics (Bjelica et al. 2024).

In Drosophila melanogaster, freezing behavior—triggered by attacks from natural enemies—was examined across approximately 100 strains, revealing that isolated individuals exhibit stronger freezing responses, whereas recovery from freezing occurs more rapidly in groups (Sato and Takahashi 2025). In *Tribolium castaneum*, individuals maintained in isolation showed a significant increase in death-feigning duration compared to pretreatment levels (Matsumura 2025). Additionally, individuals cohabiting with short-strain beetles exhibited significantly reduced death-feigning durations following cohabitation, while cohabitation with long-strain beetles did not affect synchronization of activity patterns. When a majority of individuals displayed shorter death-feigning durations, the population-level behavior synchronized towards these shorter durations (Matsumura 2025). However, no significant relationship between population density and death-feigning behavior has been reported in *T. castaneum* (Matsumura 2021).

de Lima Rocha et al. (2017) found that prior handling and habituation did not abolish the expression of tonic immobility in guinea pigs. However, habituation significantly increased the latency to initiate tonic immobility, while both handling and habituation were associated with reduced durations of tonic immobility responses. Furthermore, cohabitation with habituated individuals led to a reduction in tonic immobility duration in previously unhabituated conspecifics.

4.1.14 In a Series of Defense Actions

In dice snakes (*Natrix tessellata*), in response to predator threats, individuals that exhibit self-bleeding and defecation display shorter durations of death-feigning behavior (Bjelica and Golubović 2024). Based on these findings, Bjelica and Golubović (2024) emphasized the importance of studying sequential defensive behaviors, considering autotomy (self-bleeding), defecation, and death feigning as components of a behavioral defense sequence. Defensive behaviors in animals are highly variable, and individuals often face trade-offs between alternative antipredator strategies when confronted by a threat. For example, antlion larvae attacked by birds must make a binary decision between diving into the sand or remaining motionless. Farji-Berner et al. (2024a) found that individuals exhibiting longer diving durations into the sand had correspondingly shorter durations of death feigning.

4.1.15 Relation to Locomotor Activities

The study on dice snakes also investigated the relationship between locomotor activity and death-feigning behavior. It has been suggested that individuals inhabiting islands with few mammalian predators tend to exhibit slower escape speeds rather than death feigning, and that the intensity of death feigning is influenced by the population density of other species and allopatric populations (Bjelica et al. 2024). Edelaar et al. (2012) demonstrated intra-individual consistency in tonic immobility (TI) responses in two avian species—the yellow-crowned bishop (*Euplectes afer*) and the tree sparrow (*Passer montanus*). Individuals were exposed to both predator mounts placed near their home cages and repeated handling. The duration of tonic immobility showed significant repeatability within individuals and was negatively correlated with activity levels in the presence of the predator mount. The authors interpreted TI duration and activity rate as proxies for individual boldness, indicating a consistent behavioral syndrome across different contexts. The relationship between death-feigning behavior and locomotor activity in beetles will be discussed in detail in the following chapter.

4.1.16 Behavioral Syndromes and Animal Personality

The authors have proposed that death-feigning behavior can be considered a behavioral syndrome (Nakayama et al. 2012; Miyatake 2021). In the dung beetle species *Copris umbilicatus*, thanatosis has been characterized and analyzed as a component of behavioral personality (Natta et al. 2023). According to Natta et al. (2023), the duration of thanatosis was negatively correlated with the activity trait, such that individuals exhibiting shorter durations of thanatosis showed higher locomotor activity. No significant relationship was found between acoustic distress call traits and thanatosis duration.

More recently, death feigning has been highlighted as an indicator behavioral trait of personality in *Tribolium castaneum* (Li et al. 2025). Li et al. (2025) demonstrated that death-feigning duration, along with walking distance and emergence latency from shelters, exhibits consistent individual variation. They also found that the interaction between the duration males anticipate death and their walking distance influences mating success, with either shy but active males or bold but inactive males achieving higher numbers of mates. This suggests that personality traits, including death feigning, are linked to mating and reproductive success. This study represents the first inclusion of death feigning within the framework of animal personality research, drawing significant attention to the trait.

4.1.17 Irradiation (Radiation and Light)

No significant change in the duration of death feigning was observed in the sweet potato weevil *Cylas formicarius* following irradiation at 200 Gy for the Sterile Insect Technique (Kuriwada et al. 2010b). Hefnawy et al. (2024) investigated the effects of LED light color on behavior, including tonic immobility, using Muscovy ducks (*Cairina moschata*). They reported that tonic immobility lasted longer in individuals reared under blue LED lighting for 3 weeks compared to those reared under white, red, or yellow LEDs. Additionally, the results varied with rearing duration; at 6 and 10 weeks, tonic immobility was more frequently induced by fear under yellow and red lighting than under blue or white lighting. Saxena (1957) observed that woodlice (*Armadillidium vulgare*) exhibited stronger tonic immobility responses under reduced lighting conditions.

4.1.18 Arousal from Death Feigning

Studies have investigated the response of death-feigning adults of the vine weevil (*Otiorhynchus sulcatus*) to odors, suggesting that these individuals can be aroused from thanatosis using an olfactometer (Nakamuta et al. 2005).

Recent studies have frequently investigated factors influencing arousal from death feigning in *Tribolium castaneum*. Miyatake et al. (2019) examined the effect of vibratory stimulus intensity on arousal using an artificially selected long-duration death-feigning strain (L strain). They found that adults were successfully aroused from death feigning when the substrate (a Petri dish) was subjected to vertical vibrations of 0.2 mm amplitude. Furthermore, genetic analyses involving crosses between short-duration (S) and long-duration (L) strains, as well as F2 segregation experiments, demonstrated a genetic basis for sensitivity to vibration, which relates to the depth of death feigning. In addition, *Tribolium confusum* and *Tribolium freemani* were tested for the vibration intensity required to induce arousal from death feigning. The required vibration intensity differed among species, with *T. freemani* requiring the strongest stimulus to awaken from thanatosis (Ishihara et al. 2021).

Further studies investigating whether aggregation pheromones can stimulate arousal from death feigning in *Tribolium castaneum* demonstrated that both males and females were aroused from thanatosis upon exposure to synthetic aggregation pheromones produced by conspecific males (Ishikawa et al. 2024).

In the sweetpotato weevil *Cylas formicarius*, the effect of pheromones as a stimulus for arousal from death feigning was investigated (Himuro and Miyatake 2024). Himuro and Miyatake (2024) demonstrated that exposure to sex pheromones significantly reduced the duration of death-feigning behavior in males, whereas females were unaffected by pheromones emitted by conspecific females. This study is the first to report that sex pheromones can modulate death-feigning behavior. In addition to responses to artificially synthesized sex pheromones, experiments involving exposure to live females showed that males were immediately aroused from death feigning only in the presence of sexually mature, aged females. Males did not awaken when exposed to immature females or other males, suggesting that only sexually mature females emit effective sex pheromones (Himuro and Miyatake 2024).

4.1.19 Metabolic Rate

Metabolic rates were examined in two species, *Eucryptorrhynchus scrobiculatus* and *E. brandti*, but no correlation was found between metabolic rate and death-feigning behavior in either species (Li and Wen 2021). Similarly, in *Tribolium castaneum*, comparisons between artificially selected long- and short-duration death-feigning strains revealed no association between metabolic rate and death-feigning duration (Matsumura et al. 2025).

In contrast, Krams et al. (2013a) reported a negative relationship between resting metabolic rate (RMR) and tonic immobility duration in the yellow mealworm (*Tenebrio molitor*). Furthermore, Krams et al. (2013b) found that survival rates of *T. molitor* beetles exposed to predation by brown rats (*Rattus norvegicus*) correlated with RMR, with higher mortality observed in beetles exhibiting higher RMR. Subsequent work by Krams et al. (2014) identified a negative correlation between response latency and immobility duration. These findings demonstrate high repeatability of the complex behavioral-metabolic traits within individuals, consistent across time and environmental contexts. Specifically, beetles with higher metabolic rates required longer to enter tonic immobility upon threat and remained immobile for shorter durations. This suite of correlated traits underscores the intricate relationship between metabolism and death-feigning behavior in *T. molitor*.

4.1.20 Artificial Selection

Ten generations of artificial divergent selection for death-feigning duration in *Tribolium castaneum* successfully established strains exhibiting long (L strain) and short (S strain) durations of tonic immobility. Realized heritabilities calculated during the selection process indicated the presence of genetic variation for this trait (Miyatake et al. 2004). When predation rates were compared between these strains using the Adanson's flycatcher as a model predator, individuals from the L strain showed significantly higher survival rates than those from the S strain (Miyatake et al. 2004). However, males from the L strain, despite exhibiting better predation avoidance, achieved lower mating success due to reduced locomotor activity, suggesting a genetic trade-off between antipredator behavior and reproductive performance (Nakayama and Miyatake 2009b). A range of additional traits has also been compared between the L and S strains; these comparisons are discussed in detail in the next chapter (Fig. 4.3).

In the adzuki bean beetle (*Callosobruchus chinensis*), artificial divergent selection was conducted on two behavioral traits: the duration of death feigning and flight ability (Ohno and Miyatake 2007). After eight generations of selection, lines with longer (or shorter) durations of death feigning exhibited significantly lower (or higher) flight ability, respectively (Ohno and Miyatake 2007; Nakayama and Miyatake 2009a). Conversely, lines selected for higher (or lower) flight ability, measured by flight scores, showed shorter (or longer) durations of death feigning. These reciprocal selection experiments revealed a negative genetic correlation between death-feigning behavior and flight capacity, demonstrating a genetic trade-off between antipredator immobility and dispersal ability in this species (Ohno and Miyatake 2007).

In contrast, comparisons of life-history traits between lines subjected to artificial divergent selection for the duration of death feigning revealed that individuals from the long-duration (L) lines exhibited greater longevity, higher adult emergence rates, larger egg size, and greater reproductive effort compared to the short-duration (S) lines. However, fertility rates did not differ significantly between the two lines (Nakayama and Miyatake 2009a). Furthermore, locomotor activity was significantly lower in the S strain than in the L strain, while the number of successfully

Fig. 4.3 Relationships between temperature and frequency of death-feigning in *Callosobruchus chinensis*. Red and blue show data for females and males, respectively

mated individuals was higher in the S strain (Nakayama and Miyatake 2010). This suite of correlated traits has been collectively referred to as the "death-feigning syndrome" (Nakayama and Miyatake 2010).

4.1.21 Adaptive Significance of Death Feigning

In the salamander species *Salamandra salamandra*, juvenile larvae also exhibit death-feigning behavior as a predator avoidance strategy (Schulte and Caspers 2024). A classic study on tonic immobility in quail demonstrated that predator cats tend to target more active individuals, and that death feigning serves as an effective last-resort defense mechanism when the prey is physically attacked (Thompson et al. 1981). Another seminal work on the adaptive function of death feigning by Sargeant and Eberhardt (1975) investigated this behavior in ducks. Their study showed that feigning death increased the likelihood of escape from predation by foxes. Specifically, ducks that became motionless after being captured were sometimes able to escape, presumably because the fox relaxed its grip under the assumption that the prey was already dead.

Gallup et al. (1971) reported that juvenile chickens exhibited significantly longer durations of tonic immobility when manually restrained in the presence of a hawk model compared to those restrained without the predator stimulus. Similarly, Santos et al. (2010) found that *Liolaemus occipitalis* lizards modulate their antipredator responses based on perceived predation risk. In their study, tonic immobility induced by manual handling lasted longer when a human threat remained nearby, suggesting that the sustained presence of a predator decreases the likelihood that fleeing would be an effective survival strategy.

The larvae of *Euroleon nostras*, an antlion species, are sit-and-wait predators that construct pitfall traps in sandy substrates to ambush passing prey. Research on thanatosis in this species has been conducted under the assumption that its primary natural enemies are avian predators. Sendova-Franks et al. (2020) and Franks et al. (2021) propose the term "post-contact immobility" (PCI) as a more accurate descriptor of this behavior. These authors argue that further elaboration of immobility responses in antlion models does not confer a significant adaptive advantage, such as redirecting predator attention toward conspecifics within the same patch.

Sendova-Franks et al. (2020) and Franks et al. (2021) used modeling approaches to simulate the optimal duration of death-feigning behavior in antlions, concluding that the most advantageous strategy involves highly unpredictable durations of immobility. Consistent with this, Farji-Brener et al. (2022) demonstrated that the distribution of immobility durations follows a Poisson distribution with a mean (λ) of 1. Their findings suggest that stochastic, unpredictable durations of thanatosis enhance survival by reducing the likelihood of predator habituation or anticipation.

Death-feigning behavior has also been investigated in the horned beetle *Gnatocerus cornutus*. Matsumura et al. (2020) reported that larger males with more developed mandibles were less likely to exhibit thanatosis in response to external stimuli; however, the duration of immobility was not significantly correlated with body size. In a subsequent study, Katsuki et al. (2024) utilized strains of the giant horned owl (species name), which exhibit sex-specific predatory strategies, to demonstrate that male beetles are more likely to escape predation through active locomotion, whereas females more frequently employ tonic immobility (also referred to as freezing) as an antipredator strategy.

Skelhorn (2018) presents compelling findings using a predator–prey system involving domestic chicks (*Gallus gallus domesticus*) and stick insects. The study demonstrates that thanatosis in stick insects leads predators to misclassify live individuals as inedible or unpalatable dead prey, thereby exploiting predators' learned aversions to previously encountered unprofitable prey. Additional experimental evidence supporting the adaptive significance of death-feigning behavior, particularly using beetle models, is discussed in Chap. 5.

4.1.22 Stress/Fear

Tonic immobility in birds and mammals has been extensively studied as a behavioral indicator of fear and stress, particularly in livestock species. For example, de Haas et al. (2013) examined tonic immobility in hens as part of quality control assessments. The broader roles of stress and fear responses are further explored in Chap. 6.

4.1.23 Posture of Death Feigning

Francq (1969) conducted detailed behavioral observations of thanatosis (death feigning) in captive opossums and systematically compared it with sleep. Specifically, during sleep, the opossums close their eyes, whereas during thanatosis, the eyes remain open. The sleeping posture typically involves the dorsal side facing upward, while during thanatosis, the lateral side is oriented upward. Additionally, in sleep, the feet are tucked underneath the body, whereas in thanatosis, the feet are visible with toes usually flexed. Regarding sensory responses, sleeping individuals show no reaction to sudden sharp sounds, while during thanatosis, the ears may twitch in response to such stimuli. Upon arousal from sleep, opossums startle and turn their heads toward the source of prodding or blowing; conversely, during thanatosis, they exhibit no overt response to prodding or blowing, although slight lip retraction may occur.

4.1.24 Parasitism

The relationship between parasitism and thanatosis has also been documented. Nakao et al. (2025) demonstrated that *Tribolium confusum* adults infected with Wolbachia exhibit significantly shorter durations of death-feigning behavior compared to uninfected individuals. Additionally, sex-specific effects of parasitism on stress-induced freezing behavior have been reported in a natural beetle–nematode system.

Davis et al. (2023) observed that unparasitized female horned passalus beetles (*Odontotaenius disjunctus*) remained immobile twice as long as unparasitized males. Based on these findings, they inferred that heavily parasitized females exhibit increased boldness, whereas heavily parasitized males tend to display greater timidity.

References

Acheampong A, Mitchell BK (1997) Quiescence in the colorado potato beetle, *Leptinotarsa decemlineata*. Entomol Exp Appl 82:83–89

Akyol A, Şekeroğlu A, Tainika B, Şentürk YE, Duman M, Abacl SH, Gür FM (2024) Effects of strain and stocking density on performance, egg size distribution, egg quality, and welfare in laying hens housed in enriched cage system. Brazilian J Poultry Sci 26:1–14

Amemiya M, Sasakawa K (2021) Factors affecting thanatosis in the braconid parasitoid wasp *Heterospilus prosopidis*. Insects 12(1):48. https://doi.org/10.3390/insects12010048

Bilde T, Tuni C, Elsayed R, Pekar S, Toft S (2006) Death feigning in the face of sexual cannibalism. Biol Lett 2:23–25

Bjelica V, Golubović A (2024) Synergistic effects of musking and autohaemorrhaging on the duration of death feigning in dice snakes (*Natrix tessellata*). Biol Lett 20:20240058. https://doi.org/10.1098/rsbl.2024.0058

Bjelica V, Anđelković M, Lakušić M, Maričić M, Arsovski D, Tomović L, Golubović A (2023) A dicey situation: capture behaviours in free-ranging dice snakes. Behav Ecol Sociobiol 77:48

Bjelica V, Anđelkovic M, Marićić M, Tomović L, Bonnet X, Colubović A (2024) Tongue flicking heralds flight behaviour following passive antipredator displays in dice snakes. J Zool 322:364–374

Cardoso JCF, Mendonça JS (2019) The trade-off between fleeing and tonic immobility behaviors in an ectothermic animal. acta ethologica 22:129–134

Cassill DJ, Vo K, Becker B (2008) Young fire ant workers feign death and survive aggressive neighbors. Naturwissenschaften 95:617–624

Coutinho C, Ayres-Peres L, Araujo PB, Jara CG, Santos S (2013) Thanatosis in freshwater anomurans (Decapoda: Aeglidae). J Nat Hist 47:41–42

Davis AK, Ladd RRE, Shattuck A (2023) Sex-specific effects of a parasite on stress induced freezing behavior in a natural beetle nematode system. PLoS One 18(3):e0281149. https://doi.org/10.1371/journal.pone.0281149

Dennis DS, Lavigne RJ (1976) Ethology of *Efferia varipes* with comments on species coexistence (Diptera: Asilidae). J Kansas Entomol Soc 49:48–62

Dittrich C, Rödel MO (2023) Drop dead! Female mate avoidance in an explosively breeding frog. R Soc Open Sci. https://doi.org/10.6084/m9.figshare.c.6858610

Edelaar P, Serrano D, Carrete M et al (2012) Tonic immobility is a measure of boldness toward predators: an application of Bayesian structural equation modeling. Behav Ecol 23:619–626

Farji-Brener AG, Cajas A, Campos-Alvarado A, Castaño-Diaz M, Estévez-Haro M, Piñanez-Espejo YM (2022) Understanding unpredictability: factors influencing how long antlion larvae play dead as an antipredator behaviour. Anim Behav 194:199–204

Farji-Brener AG, Carrillo-Fajardo MY, Rodríguez-Malacara JT, Arroyo-Rodríguez V (2024a) Why are antlion larvae rare under the leaf litter? Testing the hypothesis of improper trap maintenance. Ecol Entomol 49:138–144

Farji-Brener AG, Abarca-Méndez A, Cubero-Morales L, López-Reyes K, Ramírez-Mosquera D, Escalante I (2024b) Neighbour density and post-contact immobility duration as antipredator behavior: antlion larvae do not fit the selfish prey hypothesis. Ethology 131:e13527

Farkas TE (2016) Body size, not maladaptive gene flow, explains death-feigning behaviour in *Timema cristinae* stick insects. Evol Ecol 30:623–634

Francq EN (1969) Behavioral aspects of feigned death in the opossum *Didelphis marsupialis*. Am Midl Nat 81:556–568

Franks NR, Worley A, Sendova-Franks AB (2021) Hide-and-seek strategies and post-contact immobility. Biol Lett 17:20200892

Fujioka H, Miyatake T, Ohkubo Y, Shimada M (2024) Hereditary nature of death-feigning frequency in a parasitoid wasp *Heterospilus prosopidis* using sib analysis. J Evol Biol 37:978–985

Gallup GG, Nash RF, Ellison AL (1971) Tonic immobility as a reaction to predation: artificial eyes as a fear stimulus for chickens. Psychon Sci 23:79–80

Gregory PT (2008) Bluffing and waiting: handling effects and post-release immobility in a death-feigning snake (*Natrix natrix*). Ethology 114:768–774

Gregory PT, Issac LA, Griffiths RA (2007) Death feigning by grass snakes (*Natrix natrix*) in response to handling by human "predators.". J Comp Psychol 121(2):123–129

de Haas EN, Kemp B, Bolhuis JE, Groothuis T, Rodenburg TB (2013) Fear, stress, and feather pecking in commercial white and brown laying hen parent-stock flocks and their relationships with production parameters. Poult Sci 92:2259–2269

Hefnawy E, Elgazzar E, Sabek A, El-laithy AS (2024) Effect of different LED light colors on welfare, performance, some behavioral patterns, and blood parameters of Muscovy ducks. BMC Vet Res 20:350

Hennig CW, Dunlap WP, Gallup GG (1976) The effect of distance between predator and prey and the opportunity to escape on tonic immobility in "*Anolis carolinensis*". Psychol Rec 26:313

Himuro C, Miyatake T (2024) Mate search or predation avoidance? Sex pheromone interrupts death feigning of males in the sweet potato weevil *Cylas formicarius*. J Ethol 42:169–176

Hoagland H (1927) Quantitative aspects of tonic immobility in vertebrates. Proc Natl Acad Sci USA 13:838–843

Holmes JS (1906) Death-feigning in *Ranatra*. J Comp Neurol Psychol 16:200–216

Hozumi N, Miyatake T (2005) Body-size dependent difference in death-feigning behavior of adult *Callosobruchus chinensis*. J Insect Behav 18:557–566

Igarashi K, Wada S (2024) Tonic immobility in a marine isopod: the effects of body size, sex, and colour morph. J Mar Biol Assoc UK 104:e17. https://doi.org/10.1017/S0025315424000134

Ishihara R, Mastumura K, Jones JE, Yuhao J, Fujisawa R, Nagaya N, Miyatake T (2021) Arousal from death feigning by vibrational stimuli: comparison of *Tribolium* species. J Ethol 39:107–113

Ishihara R, Matsumura K, Miyatake T (2024) Death feigning in larvae of scorpionflies (Mecoptera: Panorpidae): frequency and postural changes based on larval instars. J Insect Behav 37:152–158

Ishikawa M, Matsumura K, Miyatake T (2024) Aggregation pheromone interrupts death feigning in the red flour beetle *Tribolium castaneum*. J Ethol 42:71–75

Kang C, Moon H, Sherratt TN, Lee SI, Jablonski PG (2017) Multiple lines of anti-predator defence in the spotted lanternfly, *Lycorma delicatula* (Hemiptera: Fulgoridae). Biol J Linn Soc 120:115–124

Katsuki M, Kiyose K, Okada K (2024) Predation facilitates evolution of sex-specific antipredator defences in a sexually dimorphic beetle. Anim Behav 215:81–88

King BH, Leaich HR (2006) Variation in propensity to exhibit thanatosis in *Nasonia vitripennis* (Hymenoptera: Pteromalidae). J Insect Behav 19:241–249

Kiyotake H, Matsumoto H, Nakayama S, Sakai M, Miyatake T, Ryuda M, Hayakawa Y (2014) Gain of long tonic immobility behavioral trait causes the red flour beetle to reduce anti-stress capacity. J Insect Physiol 60:92–97

Krams I, Kivleniece I, Kuusik A et al (2013a) Predation selects for low resting metabolic rate and consistent individual differences in antipredator behavior in a beetle. Acta Ethol 16:163–172

Krams I, Kivleniece I, Kuusik A et al (2013b) Predation promotes survival of beetles with lower resting metabolic rates. Entomol Exp Appl 148:94–103

Krams I, Kivleniece I, Kuusik A, Krama T, Freeberg TM, Mänd R, Sivacova L, Rantala MJ, Mänd M (2014) High repeatability of antipredator responses and resting metabolic rate in a beetle. J Insect Behav 27:57–66

Kudo T, Hasegawa E (2022) Relationship between the duration of death-feigning and temperature in black and red elytral types of a polymorphic ladybug, *Harmonia axyridis*. J Insect Behav 35:11–16

Kuriwada T, Kumano N, Shiromoto K, Haraguchi D (2009) Copulation reduces the duration of death-feigning behaviour in the sweetpotato weevil, *Cylas formicarius*. Anim Behav 78:1145–1151

Kuriwada T, Kumano N, Shiromoto K, Haraguchi D (2010a) The effect of mass-rearing on death-feigning behaviour in the sweet potato weevil (Coleoptera: Brentidae). J Appl Entomol 134:652–658

Kuriwada T, Kumano T, Shiromoto N, Haraguchi D (2010b) Effect of irradiation on death-feigning behavior in the male sweetpotato weevil *Cylas formicarius* (Coleoptera: Blentidae). Fla Entomol 93:39–44

Kuriwada T, Kumano N, Shiromoto K, Haraguchi D (2011) Age-dependent investment in death-feigning behaviour in the sweetpotato weevil *Cylas formicarius*. Physiol Entomol 36:149–154

Lawrence SE (1992) Sexual cannibalism in the praying mantid, Mantis religiosa: a field study. Anim Behav 43:569–583

Li H, Wen J (2021) Factors influencing the duration of death feigning in *Eucryptorrhynchus scrobiculatus* and *E. brandti* (Coleoptera: Curculionidae). J Ethol 40:61–69

Li H, Zhang G, Ji Y, Wen J (2018) Effects of starvation on death-feigning in adult *Eucryptorrhynchus brandti* (Coleoptera: Curculionidae). Ethology 125:645–651

Li R, Yang C, Lou Y, Lü J, Lyu N (2025) Animal personalities influence male mating behavior and reproductive fitness. Behav Ecol 36(1):arae094

de Lima Rocha AD, Menescal-de-Oliveira L, da Silva LFS (2017) Effects of human contact and intra-specific social learning on tonic immobility in guinea pigs, *Cavia porcellus*. Appl Anim Behav Sci 191:1–4

Lira AFA, Almedia FM, Alburquerque CMR (2020) Reaction under the risk of predation: effects of age and sexual plasticity on defensive behavior in scorpion Tityus pusillus (Scorpiones: Buthidae). J Ethol 38:13–19

Matsumura K (2021) Relationship between death-feigning behavior and population density in a beetle. J Ethol 39:363–367

Matsumura K (2025) Synchronising anti-predator behaviour in the red flour beetle *Tribolium castaneum*. Behav Ecol 36:araf013

Matsumura K, Yumise K, Fujii Y, Hayashi T, Miyatake T (2020) Anti-predator behaviour depends on male weapon size. Biol Lett 16:20200601

Matsumura K, Hosken D, Noda T, Miyatake T, Sharma MD (2025) Behavior, behavioral syndromes, and metabolism: the effects of artificial selection for death-feigning on metabolic rate. J Insect Sci 25:16; ieaf007

Mignon-Grasteau S, Chantry-Darmon C, Boscher M-Y, Sellier N, Le Bihan-Duval E, Bertin A (2017) Genetic determinism of fearfulness, general activity and feeding behavior in chickens and its relationship with digestive efficiency. Behav Genet 47:114–124

Miyatake T (2001a) Diurnal periodicity of death-feigning in *Cylas formicarius* (Coleoptera: Brentidae). J Insect Behav 14:421–432

Miyatake T (2001b) Effects of starvation on death-feigning in adults of *Cylas formicarius* (Coleoptera: Brenridae). Ann Entomol Soc Am 94:612–616

Miyatake T (2021) Environmental, physiological, and genetic effects on tonic immobility in beetles. In: Sakai M (ed) Death-feigning in insects, entomology monographs. Springer Nature Singapore Pte Ltd, pp 39–53

Miyatake T, Katayama K, Takeda Y, Nakashima A, Sugita A, Mizumoto M (2004) Is death-feigning adaptive? Heritable variation in fitness difference of death-feigning behaviour. Proc R Soc Lond Ser B 271:2293–2296

Miyatake T, Okada K, Harano T (2008) Negative relationship between ambient temperature and death-feigning intensity in adult *Callosobruchus maculatus* and *Callosobruchus chinensis*. Physiol Entomol 33:83–88

Miyatake T, Matsumura K, Kitayama R, Otsuki K, Yuhao J, Fujisawa R, Nagaya N (2019) Arousal from tonic immobility by vibration stimulus. Behav Genet 49:478–483

Moore AU, Amstey MS (1962) Tonic immobility: differences in susceptibility of experimental and normal sheep and goats. Science 135:729–730

Nakamuta K, van Tol WHM, Visser JH (2005) An olfactometer for analyzing olfactory responses of death-feigning insects. Appl Entomol Zool 40:173–175

Nakao K, Sumi T, Matsumura K (2025) Antibiotic treatment eliminated *Wolbachia* in *Tribolium confusum* (Coleoptera: Tenebrionidae) and prolonged tonic immobility duration in later generations. https://doi.org/10.1007/s13355-025-00915-x

Nakayama S, Miyatake T (2009a) Positive genetic correlations between life-history traits and death-feigning behavior in adzuki bean beetle (*Callosobruchus chinensis*). Evol Ecol 23:711–722

Nakayama S, Miyatake T (2009b) Genetic trade-off between abilities to avoid attack and to mate: a cost of tonic immobility. Biol Lett 6:18–20

Nakayama S, Miyatake T (2010) A behavioral syndrome in the adzuki bean beetle: genetic correlation among death feigning, activity, and mating behavior. Ethology 116:108–112

Nakayama S, Sasaki K, Matsumura K, Lewis Z, Miyatake T (2012) Dopaminergic system as the mechanism underlying personality in a beetle. J Insect Physiol 58:750–755

Natta G, Laini A, Roggero A, Fabbriciani F, Rolando A, Palestrini C (2023) Behavioural repeatability and behavioural syndrome in the dung beetle *Copris umbilicatus* (Coleoptera, Scarabaeidae). Insects 14:529. https://doi.org/10.3390/insects14060529

O'Brien TJ, Dunlap WP (1975) Tonic immobility in the blue crab (*Callinectes sapidus*, Rathbun) its relation to threat of predation. J Comp Physiol Psychol 89:86–94

Odén K, Gunnarsson S, Berg C, Algers B (2005) Effects of sex composition on fear measured as tonic immobility and vigilance behaviour in large flocks of laying hens. Appl Anim Behav Sci 95:89–102

Ohno T, Miyatake T (2007) Drop or fly? Negative genetic correlation between death-feigning intensity and flying ability as alternative anti-predator strategies. Proc R Soc B 274:555–560

Ouyang H, Wu P, Zhang R, Haseeb M (2020) Trade-off investment between tonic immobility and mate search in the Sweetpotato weevil, *Cylas formicarius* (Coleoptera: Brentidae). Insects 11:774. https://doi.org/10.3390/insects11110774

Ouyang H, Zhang R, Haseeb M (2023) Investment trade-off between mating behavior and tonic immobility in the Sweetpotato weevil *Cylas formicarius* (Coleoptera: Brentidae). Insects 14:73. https://doi.org/10.3390/insects14010073

Oyen KJ, Croucher L, Benoit JB (2021) Tonic immobility is influenced by starvation, life stage, and body mass in ixodid ticks. J Med Entomol 58:1030–1040

Pereyra P, Saraco M, Maldonado H (1999) Decreased response or alternative defensive strategies in escape: two different types of long-term memory in the crab *Chasmagnathus*. J Comp Physiol A 184:301–310

Quadros AF, Bugs PS, Araujo PB (2012) Tonic immobility in terrestrial isopods: intraspecific and interspecific variability. Zookeys 20:155–170

Santos MB, Oliveira MCLM, Verrastro L, Tozetti AM (2010) Playing dead to stay alive: death-feigning in *Liolaemus occipitalis* (Squamata: Liolaemidae). Biota Neotropica 10:361–364

Sargeant AB, Eberhardt LE (1975) Death feigning by ducks in response to predation by red foxes (*Vulpes fulva*). Am Midl Nat 94:108–119

Sato DX, Takahashi Y (2025) Neurogenomic and behavioral principles shape freezing dynamics and synergistic performance in *Drosophila melanogaster*. Nat Commun 6:5928. https://doi.org/10.1038/s41467-025-61313-z

Saunders DS (2002) Insect clocks, 1st edn. Elsevier Science, p 576

Saxena SC (1957) An experimental study of thanatosis in *Armadillidium vulgare* (Latreille). J Zool Soc India 9:192–199

Schulte L, Caspers BA (2024) Tonic immobility behaviour does not differ between fire salamander larvae from ponds and streams. Ecol Evol 14:e11211

Segovia JMG, Moura RR, Willemart RH (2019a) Starvation decreases behavioral consistency in a Neotropical harvestman. Acta ethologica 22:203–208

Segovia JMG, Murayama GP, Willemari RH (2019b) Sexual differences in weaponry and defensive behavior in a neotropical harvestman. Curr Zool 65:553–558

Sendova-Franks AB, Worley A, Franks NR (2020) Post-contact immobility and half-lives that save lives. Proc R Soc B 287:20200881. https://doi.org/10.1098/rspb.2020.0881

Skelhorn J (2018) Avoiding death by feigning death. Curr Biol 28:R1121–R1142

Suzuki S, Futami K (2017) Predatory risk increased due to egg-brooding in *Armadillidium vulgare* (Isopoda: Oniscidea). Ethology 124:256–259

Suzuki K, Ikebuchi M, Okanoya K (2013) The impact of domestication on fearfulness: a comparison of tonic immobility reactions in wild and domesticated finches. Behav Process 100:58–63

Taylor KL, Henry CS, Farkas TE (2023) Why fake death? Environmental and genetic control of tonic immobility in larval lacewings (Neuroptera: Chrysopidae). J Insect Sci 23:1–5

Thompson RKR, Foltin RW, Sweet BA, Graves CA, Lowitz CE (1981) Tonic immobility in Japanese quail can reduce the probability of sustained attack by cats. Anim Lean Behav 9:145–149

Tojo S (1991) Variation in phase polymorphism in the common cutworm, *Spodoptera Iitura* (Lepidoptera:Noctuidae). Appl Entomol Zool 26:571–578

Death-Feigning Syndrome 5

Abstract

In this chapter, I will discuss the background that led me to begin artificial selection and evolutionary experiments on death-feigning behavior, which I have been conducting since 1997. I will summarize the materials and methods used, as well as the results and their interpretation. To better understand the genetic basis of death feigning, the breeding experiments not only focused on direct responses but also examined correlated responses in various traits. In studies on the *Tribolium* beetle, working alongside a physiologist, we investigated the dynamics of biogenic amines inside the bodies of beetles exhibiting death-feigning behavior, pursuing the physiology underlying this response. At the core of this work was dopamine, a key neurotransmitter that governs movement in animals. Furthermore, with new collaborators in molecular biology, we applied molecular techniques to explore the mechanisms controlling the duration of death feigning. Together with researchers in engineering, we also conducted experiments that revealed links between death feigning and gait abnormalities in beetles. These findings suggest potential parallels between death-feigning behavior and human motor disorders. We propose that this constellation of traits constitutes what we call the "death-feigning syndrome."

Keywords

Artificial selection · Correlated response · Heritability · Experimental evolution · Molecular · Dopamine · Transcriptome analysis · Insect trajectory · Parkinson's disease

---

Supplementary Information The online version contains supplementary material available at https://doi.org/10.1007/978-981-95-5319-8_5.

© The Author(s), under exclusive license to Springer Nature Singapore Pte Ltd. 2026

T. Miyatake, *Death Feigning*, Ecological Research Monographs, https://doi.org/10.1007/978-981-95-5319-8_5

5.1 Brief Introduction

Artificial selection experiments represent one of the most powerful quantitative genetic approaches for elucidating the genetic architecture of biological traits, including correlated responses, variation in physiological characteristics, and underlying genomic factors (Falconer 1989). In this context, artificial selection has been employed to manipulate the duration of death feigning—both in the direction of increased and decreased duration—in several beetle species. These include the red flour beetle (*Tribolium castaneum*) (Miyatake et al. 2004, 2008, 2009; Matsumura and Miyatake 2018), the confused flour beetle (*T. confusum*) (Nakayama and Miyatake 2010), the Kashmir flour beetle (*T. freemani*) (Konishi et al. 2020), the adzuki bean weevil (*Callosobruchus chinensis*) (Ohno and Miyatake 2007), and the sweetpotato weevil (*Cylas formicarius*) (Ouyang et al. 2020, 2023). In all of these studies, divergent selection lines were successfully established: long-duration (L) strains selected for prolonged death-feigning, and short-duration (S) strains selected for abbreviated responses. The consistent increase in phenotypic divergence across successive generations observed in all species strongly indicates that death-feigning duration is a quantitative trait. Furthermore, the retention of additive genetic variance across generations implies that selection responses are both heritable and repeatable. The subsequent section outlines the species-specific results of these artificial selection experiments, including their correlated responses. An earlier version of this chapter was originally published in Japanese in 2022 under the title "*'Shinda furi' de ikinobiru*" by Iwanami Shoten, Tokyo. Reproduced with permission from Iwanami Shoten, Publishers.

5.2 The Results of *Cylas formicarius*

I initiated research on death-feigning behavior in the sweetpotato weevil (*Cylas formicarius*) in the 1990s and was appointed Associate Professor at Okayama University in the autumn of 2000. Although *C. formicarius* is widely distributed across the Southwestern Islands of Japan, it does not occur naturally north of Kyushu, including Okayama Prefecture. Consequently, under Japan's Plant Protection Law, the importation of this pest species from southern regions to Okayama was subject to strict regulatory controls (Moriya and Miyatake 2001). While special permission from the Ministry of Agriculture, Forestry and Fisheries (MAFF) would have allowed for the transportation and containment of the weevil within a designated high-security facility, the potential risks remained non-negligible. Given that university laboratories are accessible to students and not fully closed systems, the possibility of accidental escape posed a serious biosecurity concern. This was especially critical considering the regional economic importance of sweetpotato agriculture in Okayama and neighboring Shikoku. A single containment failure could lead to significant agricultural damage and public outcry. To mitigate these risks and to ensure both scientific rigor and institutional safety, the author elected to discontinue the use of *C. formicarius* and instead adopt alternative model organisms for behavioral genetic studies on death feigning. Specifically, *Tribolium* species and the adzuki bean beetle (*Callosobruchus chinensis*) were selected as surrogate species for conducting artificial selection experiments targeting variation in death-feigning duration.

5.3 Artificial Selection Experiment with *Tribolium* Beetles

Three primary factors motivated the selection of *Tribolium* spp. (flour beetles) and *Callosobruchus chinensis* (adzuki bean beetle) as substitute model organisms for death-feigning research.

First, these species are highly amenable to laboratory rearing due to their minimal maintenance requirements and low associated costs. *C. chinensis* requires only adzuki beans, while *Tribolium* species thrive on flour. Additionally, the laboratory had preexisting infrastructure and equipment optimized for handling stored-product pests, facilitating a smooth transition to working with these beetles (Fig. 5.1).

Second, these organisms are particularly suitable for undergraduate education and long-term student-led research projects. Even with minimal intervention, stable populations can be maintained as long as food is available, making them ideal for

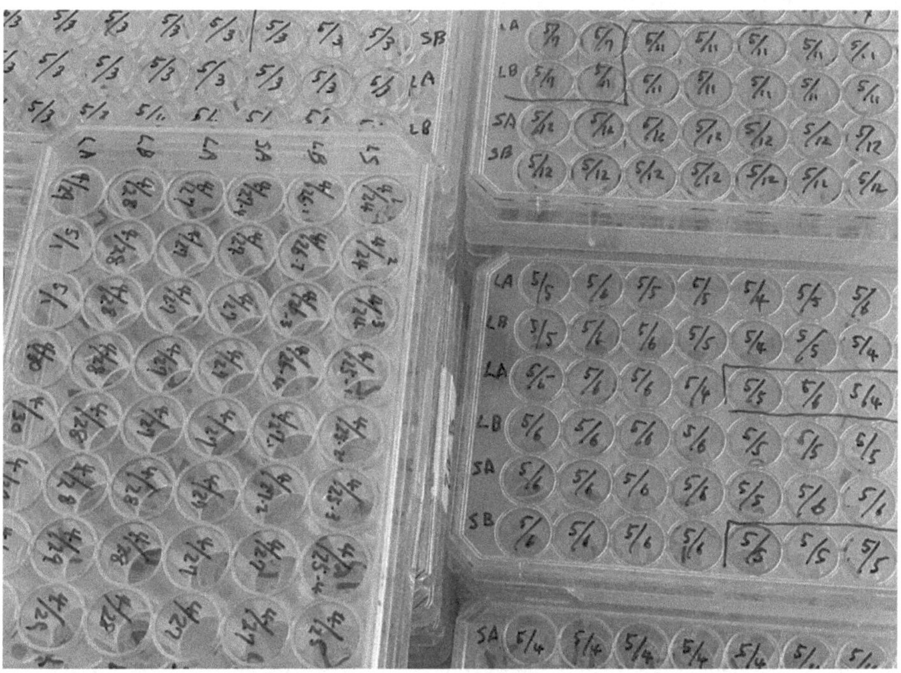

Fig. 5.1 *Tribolium* beetles can be individually reared in the multi-well containers shown in this photograph. By marking the eclosion date on each well with a permanent marker, a large number of individuals can be managed and monitored individually in a space-efficient manner

students with varying levels of experience in insect husbandry. Since individual aptitude for maintaining live insect colonies is not immediately apparent, it is advantageous to use robust species that tolerate a range of care quality. Unlike many Lepidoptera, these beetles exhibit a low incidence of viral or fungal outbreaks under standard rearing conditions. Although occasional mite infestations can occur, they are manageable with routine care.

Third—and most critically for evolutionary research—both species have short generation times (approximately 1–2 months), enabling the observation of heritable trait changes across multiple generations within a single academic year. This allows for the empirical tracking of evolutionary dynamics, such as changes in death-feigning behavior under artificial selection, within the timeframe of a typical undergraduate thesis. The capacity to observe evolutionary change in real time makes these beetles excellent educational and research models for experimental evolutionary biology.

A third and crucial reason for selecting these two species was that both *Tribolium* and *Callosobruchus* exhibit death-feigning behavior, making them directly relevant to the core research question. Moreover, at the time of selection in the early 2000s, genomic resources for *Tribolium castaneum* were rapidly advancing. Preliminary information regarding genome sequencing efforts in the United States suggested that a full genome assembly was forthcoming. This presented a unique opportunity: if genetically divergent lines for death feigning could be established through artificial selection within a few years, they could later serve as powerful tools for identifying the genetic architecture underlying this behavioral trait. Indeed, by 2008, the *T. castaneum* genome had been fully sequenced and made publicly available (*Tribolium* Genome Sequencing Consortium 2008).

Beyond their utility for research on antipredator behavior, both beetle species also provided excellent systems for student-led projects on mating behavior and reproductive strategies. These included undergraduate graduation theses and doctoral dissertations (e.g., Yanagi and Miyatake 2002, 2003; Miyatake and Matsumura 2004; Yamane and Miyatake 2005; Katsuki and Miyatake 2009; Yuhao et al. 2021). Thus, the use of *T. castaneum* and *C. chinensis* supported a broad array of educational and research objectives, ranging from behavioral ecology to evolutionary genetics.

This section details the methodology used for artificial selection on death-feigning duration in *Tribolium castaneum*. Experimental individuals were derived from a stock population maintained at the University of Tsukuba. For each generation, 100 unmated males and 100 unmated females, approximately 2 weeks post-eclosion, were randomly selected from newly emerged adults to constitute the breeding cohort for the next generation. To standardize behavioral measurements, individuals were isolated 24 hours prior to testing by placing them in separate wells of 24-well plates. This isolation period was implemented to minimize tactile stimulation from conspecifics, which has been shown to significantly influence the duration of tonic immobility. Previous studies in other species, such as the Colorado potato beetle (*Leptinotarsa decemlineata*), demonstrated that prior physical contact or stimulation can alter subsequent death-feigning responses (Acheampong and

Mitchell 1997). Our preliminary observations also indicated that such stimulation may either suppress or potentiate immobility behavior depending on the individual, further justifying the need for standardized pretest isolation.

On the day of behavioral assessment, beetles were individually transferred to a white ceramic dish and subjected to a standardized stimulus: each individual was picked up using forceps and dropped from a height of 5 cm. The duration of the resulting death-feigning response (tonic immobility) was recorded. Only the initial response was measured to avoid confounding effects of repeated stimulation, which our experience has shown may lead to behavioral plasticity—some individuals ceasing to feign death, while others exhibit sensitization and increasingly prolonged responses.

To quantify death-feigning behavior, a stopwatch was used to measure the duration of tonic immobility in 100 males and 100 females subjected to standardized stimulation. If an individual did not exhibit tonic immobility upon the initial stimulus, the stimulus was repeated. Individuals that failed to respond after three consecutive stimulations were recorded as exhibiting zero seconds of death-feigning duration. Based on the recorded durations, the 10 males and 10 females with the longest immobility times were selected and allowed to mate randomly. Their offspring constituted the "long" (L) strain. Conversely, the 10 males and 10 females with the shortest immobility durations were selected and randomly mated to produce the "short" (S) strain. This bidirectional selection procedure was repeated for 10 successive generations to establish divergent lines exhibiting consistently long or short durations of death-feigning behavior.

The methods and results of the 10-generation artificial divergent selection experiment are detailed in Miyatake et al. (2004). The selection procedure yielded a clear and robust response. In the base population, the majority of individuals exhibited tonic immobility for approximately 10 seconds upon stimulation. However, after 10 generations of directional selection for increased immobility, the derived population consistently displayed death-feigning durations exceeding 2 minutes. This population is hereafter referred to as the "long" (L) strain. Conversely, in the line selected for reduced death-feigning duration, individuals exhibited a marked decline in immobility behavior. By the tenth generation, most individuals failed to show any tonic immobility, regardless of repeated stimulation. This population is hereafter referred to as the "short" (S) strain. To confirm that the observed responses were due to selection rather than stochastic processes, the entire selection experiment was independently replicated. Without replication, there remains a nontrivial probability that observed phenotypic divergence could arise from random changes in allele frequencies—an evolutionary process known as genetic drift (Falconer 1989). Replication thus ensured that the divergent responses were reproducible and attributable to the imposed selective pressures rather than chance.

After just 10 generations of artificial selection, striking divergence was observed in the proportion of individuals exhibiting tonic immobility. In the long strain, nearly 100% of individuals consistently feigned death in response to stimulation. In contrast, the majority of individuals in the short strain no longer exhibited any death-feigning behavior (Miyatake et al. 2004). This phenotypic divergence highlights the existence of individual-level variation in death-feigning behavior, and more importantly, confirms that such behavioral differences are heritable. The successful establishment of these divergent strains demonstrates that death-feigning behavior can respond to selection, fulfilling two key conditions required for evolution by natural selection: the presence of heritable variation and differential expression of the trait. However, a third condition remains essential for confirming evolutionary potential—evidence that the trait is subject to selective pressure. Specifically, it is necessary to assess whether individuals who feign death experience differential survival or reproductive success compared to those who do not. Determining the fitness consequences of tonic immobility is thus critical for evaluating its adaptive significance.

5.4 Is Feigning Death Adaptive?

Given that the artificial selection experiments are based on the assumption that death-feigning behavior has evolved as an antipredator adaptation, it is essential to demonstrate that the two strains—long and short—differ in their susceptibility to predation. Only by showing a significant difference in predation rates between these strains can death feigning be confirmed as an effective survival strategy. In other words, the evolutionary advantage of tonic immobility must be validated through evidence of differential fitness under ecologically relevant conditions.

It is also appropriate in this context to refer to death feigning as a "strategy" rather than a "tactic." In evolutionary biology, a *strategy* refers to a behavior that is genetically encoded and transmitted across generations, reflecting long-term evolutionary optimization. In contrast, a *tactic* denotes a flexible behavioral response employed by individuals in reaction to specific environmental or situational cues, without necessarily having a heritable basis. Thus, once heritability has been established—as it has been in the present study—death feigning can be appropriately described as a genetically based behavioral strategy.

At this stage, a fundamental problem emerged: *What is the natural enemy of Tribolium castaneum?* This question became a topic of active discussion among the students in my laboratory, many of whom had a deep interest in animals. One student jokingly proposed that his pet turtles might serve as suitable predators. However, such an interaction is highly unlikely to occur under natural conditions, as turtles and *Tribolium* beetles do not share the same habitat. For the predator–prey relationship to be ecologically meaningful, it must involve species that coexist in the same environment. To address this, I visited several rice and flour mills—habitats where *T. castaneum* is commonly found—to observe what kinds of organisms coexist with the beetles. During these site visits, I discovered that certain indoor spiders, particularly jumping spiders (family Salticidae), were frequently found in the same spaces as *T. castaneum*. This raised a promising possibility: could these spiders serve as natural predators of the beetle? The next question, then, was: *Which species of jumping spider actually preys upon Tribolium beetles in these environments?*

To empirically assess the adaptive value of death-feigning behavior in *Tribolium castaneum*, it was necessary to identify a natural predator that coexists with the beetle in its typical habitat. A student with expertise in spiders investigated this question as part of an undergraduate thesis. Initial trials revealed that most grassland-dwelling jumping spiders (Salticidae) did not attack or consume adult *T. castaneum*. However, one species frequently found indoors—*Adanson's house jumper* (*Hasarius adansoni*)—did actively prey on adult beetles. This observation was later substantiated in a detailed study (Hayashi et al. 2024).

Further investigations confirmed that two species of synanthropic jumping spiders commonly found in indoor environments—*Hasarius adansoni* (Audouin, 1826) and *Plexippus paykulli* (Audouin, 1826)—preyed upon adult *T. castaneum*, but in a size-dependent manner. Larger individuals, particularly adult females, exhibited a higher probability of successful predation (Hayashi et al. 2024). In contrast, three species of grassland-associated jumping spiders—*Carrhotus xanthogramma* (Latreille), *Mendoza elongata* (Karsch), and *Myrmarachne elongata* Szombathy—did not attack or feed on adult *T. castaneum* under experimental conditions. Interestingly, all five of these salticid species—including both indoor and grassland forms—readily preyed upon *T. castaneum* larvae (Hayashi et al. 2024), suggesting that larval stages are more vulnerable across a broader range of spider species.

Based on these findings, and after confirming the presence of *Hasarius adansoni* in a rice mill inhabited by *T. castaneum*, I selected this species as the model predator for subsequent behavioral and ecological assays. The search for a suitable predator, specifically *Hasarius adansoni* (Adanson's house jumper), was initiated in earnest. With the cooperation of a research colleague at a nearby university, we successfully collected over 30 adult female individuals for use in predation experiments. Female spiders were specifically targeted for collection due to notable sexual size dimorphism in *H. adansoni*; males are significantly smaller in body size than females. Previous observations, later confirmed in detail by Hayashi et al. (2024), demonstrated that smaller-bodied individuals, including many males, often failed to prey upon adult *T. castaneum*. In contrast, females—due to their larger size—consistently exhibited successful predation on adult beetles. Thus, the exclusive use of females ensured more reliable and ecologically relevant predation outcomes in subsequent behavioral assays.

5.5 A World First: Experimental Evidence That Death-Feigning Behavior Confers Antipredator Advantage

Predation trials were conducted by placing a single *T. castaneum* adult—either from the long or short strain—together with a single adult female *Hasarius adansoni* (collected from the field or provided by collaborators) in a Petri dish. The spider's behavior was then observed under controlled laboratory conditions. Interestingly, a consistent pattern emerged: upon initiating an attack, *H. adansoni* often released the beetle shortly after physical contact, rather than proceeding directly to consumption. This response contrasts with the spider's typical predation behavior toward soft-bodied prey such as flies, which are usually subdued and consumed immediately following capture. One plausible explanation for this hesitation is the beetle's hard exoskeleton. Upon biting the beetle, the spider appeared to be startled—possibly due to unexpected resistance or tactile feedback—and subsequently released its prey. This "startle and release" response was repeatedly observed across trials, suggesting that the beetle's morphology, in combination with its death-feigning behavior, may play a synergistic role in predator deterrence.

The behavior exhibited by the prey at the moment of predatory attack proved to be a critical determinant of survival. When attacked by *Hasarius adansoni*, beetles from the long strain frequently entered a state of tonic immobility in response to the stimulus.

The long-strain individuals feigned death in 12 out of 14 trials (85.7%) following a spider attack. In most cases, the spider approached and visually inspected the immobile beetle for a short period, but eventually lost interest and disengaged. As a result, 13 of the 14 beetles (92.9%) survived the predation attempt, suggesting that tonic immobility effectively deterred further attacks.

In contrast, the short strain beetles, which had been selected for minimal death-feigning behavior, exhibited a markedly different pattern. In 13 of 14 trials (92.9%), individuals failed to feign death when attacked, despite the intensity of the stimulus. Interestingly, even individuals from the short strain—typically unresponsive to repeated mechanical stimulation—were visibly startled by the spider's attack, yet only one individual exhibited tonic immobility. This sole immobile individual survived, while the remaining 13, which responded by jittering or walking within the spider's visual field, were rapidly re-detected, attacked, and ultimately consumed—often within two or three strikes. These observations provide compelling evidence that tonic immobility acts as an effective antipredator defense, substantially increasing the probability of survival when facing visually hunting predators such as jumping spiders.

In quantitative terms, 13 out of 14 individuals (92.9%) from the long strain survived predation attempts by *Hasarius adansoni*, while only 5 out of 14 individuals (35.7%) from the short strain survived. Statistical analysis revealed that the probability of observing such a difference by chance alone is 0.16% ($p = 0.0016$), indicating a highly significant effect of strain on survival outcome (Miyatake et al. 2004). This result provides strong empirical support that tonic immobility significantly enhances survival when beetles are exposed to visually hunting predators.

This experiment represents the first statistically robust, quantitative demonstration that death-feigning behavior serves as an effective antipredator strategy. The substantial difference in survival between strains subjected to divergent selection confirms that death feigning is not merely a reactive behavior, but an *adaptive trait* subject to *natural selection*. The findings fulfill all three key criteria for evolutionary adaptation: genetic variation, heritability, and differential fitness.

These results were published in the *Proceedings of the Royal Society B: Biological Sciences* under the title "Is death-feigning adaptive? Heritable variation in fitness-related traits in a beetle" (Miyatake et al. 2004), marking a milestone in the scientific validation of tonic immobility as an evolved behavioral strategy.

5.6 Selfish Foraging Strategies: Adaptive Behaviors That Promote Individual Survival Through the Exploitation of Conspecifics

While death feigning has been shown to confer survival benefits for solitary *Tribolium castaneum*, individuals in natural populations generally inhabit aggregations. Considering the social context of these beetles uncovers further adaptive functions of death-feigning behavior. The adaptive significance of death-feigning behavior was first documented in a study published in *Proceedings of the Royal Society B* in November 2004. Following this, on April 13, 2006, the UK journal *Nature* featured a short commentary titled "Grasshoppers Do Not Play Dead" (Ruxton 2006).

Consistent with its title, the *Nature* article focused on the behavior of Acrididae grasshoppers exhibiting rigidity of the legs along the longitudinal axis and remaining immobile. The article also noted that death feigning may function as a physical antipredator strategy by prompting predators to regurgitate captured individuals. In the spiny Acrididae grasshopper, ingestion by a frog triggers the vertical extension of the spiny hind legs. Consequently, the thoracic spines, naturally present in adult grasshoppers, protrude laterally, while the vertically erected hind legs extend longitudinally, resulting in a rigid, thistle-like posture, which gives the species its common name. Frogs are unable to swallow grasshoppers in this stiffened posture. This phenomenon was documented in a study published earlier that year by a graduate student at Kyoto University.

Importantly, spiny Acrididae grasshoppers do not assume this posture in response to avian attacks; the rigid immobility is triggered solely upon ingestion by frogs. Hence, this death-feigning behavior operates as a targeted physical defense to avoid predation by frogs. The *Nature* commentary emphasized that death feigning may have adaptive roles beyond simply rendering prey unrecognizable to predators, illustrating that the behavior can serve distinct functional purposes depending on the predator context.

Within the article, our 2004 study was highlighted, in which we showed that death-feigning behavior in *Tribolium castaneum* confers an adaptive advantage against predation by jumping spiders (Salticidae). Importantly, the commentary proposed that, because *Tribolium castaneum* secretes a noxious chemical, the observed immobility may extend beyond a conventional death-feigning signal. The beetle may intentionally assume a supine posture to advertise its chemical defenses to potential predators, thereby transmitting a potent signal of unpalatability. This perspective challenges the traditional interpretation of death feigning as a purely defensive mechanism, suggesting instead that the behavior, although manifesting as tonic immobility, may function offensively to communicate toxicity rather than solely to diminish predator interest.

The commentary further proposed that should the hypothesis—that the supine death-feigning posture serves as a chemical warning—be valid, one would expect selectively bred long-duration death-feigning lines of *Tribolium castaneum* to possess higher concentrations of defensive compounds compared to short-duration, unselected lines. This article was authored by Professor Graham R. Luxton, who previously authored *Predator Avoidance: The Evolutionary Ecology of Crypsis, Warning, and Mimicry*.

The significance of this commentary was immediately apparent. *Nature* is a journal with a global readership among scientists, and the article not only highlighted our previous work but also proposed that our 2004 study, which established the effectiveness of death feigning as an antipredator strategy, may have overlooked additional adaptive functions. Crucially, death feigning in *Tribolium castaneum* involves more than a stereotyped period of immobility in response to external stimuli; it may serve as a communicative posture signaling unpalatability to predators. Implicitly, the commentary encouraged further empirical testing of this hypothesis, which I interpreted, in a constructive sense, as a scientific challenge.

5.7 Is Death Feigning a Posture That Advertises Unpalatability?

As a matter of scientific integrity, I regarded it essential to respond to any challenge posed. It is important to note that the intent of such a challenge is not to undermine the commentator, but to encourage rigorous inquiry and advance constructive scientific dialogue. This commentary reignited my scientific curiosity at a time when I had felt a certain sense of burnout, believing that the adaptive significance of death feigning had already been clarified and that my work was essentially complete. If such a hypothesis is proposed, then it is worth investigating. While perhaps somewhat ambitious, I was genuinely pleased to receive this commentary, as it provided a compelling impetus for further research.

Tribolium castaneum is documented to contain methyl-1,4-benzoquinone (MBQ, 98% purity), a bitter-tasting compound with recognized chemical defense properties. Application of mechanical stress, such as grasping, elicits the emission of a strongly odorous secretion, which is expected to serve as a deterrent to natural predators, including jumping spiders. The possession of this chemical substance by *Tribolium castaneum* has been proposed to function as an external defense mechanism against predation. Lacking formal training in chemistry, I was initially unsure how to address this experimentally. Fortunately, I was affiliated with the Faculty of Agriculture at Okayama University, which houses an Agricultural Chemistry division, including faculty specializing in insect chemical ecology. Upon consulting a professor with whom I had an established professional rapport, I inquired about the feasibility of quantifying benzoquinone in the beetles. The professor confirmed that such measurements could be readily performed.

I promptly discussed the potential significance of this study with a graduate student in our laboratory, who expressed interest. Under the supervision of a professor specializing in Agricultural Chemistry, we utilized gas chromatography to compare benzoquinone concentrations between short- and long-duration death-feigning lines of *Tribolium castaneum*. A total of 2020 adult individuals from each line were analyzed to determine their relative benzoquinone content. No significant inter-line differences were detected, thereby falsifying Professor Ruxton's prediction that long-duration lines would exhibit higher internal benzoquinone levels.

5.8 The Behavioral Response Upon Predatory Attack Is Critical

Importantly, the relevant measure may not be the baseline internal benzoquinone content, but the quantity emitted by *Tribolium castaneum* during actual predator encounters. This prompted the hypothesis that long-duration death-feigning lines may release higher amounts of benzoquinone upon attack. Despite initial concerns regarding the feasibility of quantifying volatile chemical emissions, I sought guidance once more from the professor specializing in Agricultural Chemistry, Prof. Shuhei Nakajima.

The professor indicated that such measurements were feasible, which impressed me greatly and underscored the advantages of a multidisciplinary research environment such as the Faculty of Agriculture. Guided by this expertise, we initiated a collaborative study. Regular collection of female Adanson's jumping spiders was undertaken to ensure sufficient experimental subjects.

Experimental procedures were as follows: adult *Tribolium castaneum* and Adanson's jumping spiders were co-introduced into cylindrical glass vials (12 mm diameter × 40 mm height). The vials were sealed with Parafilm, and upon initiation of predation by the spider, a single adsorption fiber capable of capturing airborne chemical compounds was inserted into the vial. Airborne volatiles were adsorbed for 5 minutes and subsequently quantified using gas chromatography, allowing comparison of benzoquinone emission between long- and short-duration death-feigning lines during predator encounters.

Notably, the experiments revealed that *Tribolium castaneum* did not release the bitter compound benzoquinone when escaping predation through death-feigning behavior in response to jumping spider attacks. This result was consistent for both short- and long-duration death-feigning lines. Benzoquinone emission was observed only when beetles were repeatedly attacked and ultimately consumed by spiders, indicating that this compound does not function in deterring initial predation attempts.

Reexamination of prior studies indicates that benzoquinone emitted by *T. castaneum* exhibits insecticidal properties against cohabiting larvae of moths and flies in enclosed environments. However, the adaptive significance of benzoquinone released postmortem remains unresolved, and further research is required to elucidate its role.

The experimental results effectively falsified both Professor Ruxton's challenge and my personal interpretation of it. Nevertheless, the prospect of publishing a manuscript solely to report that the challenge had been negated seemed insufficient. I considered whether it might be possible to integrate an exploration of the true adaptive significance of death feigning in *Tribolium castaneum* alongside the experimental refutation, thereby providing a more complete and scientifically meaningful response to Professor Luxton's inquiry. In the course of these reflections, certain unresolved issues within our 2004 dataset gradually manifested as persistent conceptual ambiguities.

It is worth noting that, in prior experiments, 5 out of 14 individuals in the short-duration line survived predation, compared with 13 out of 14 in the long-duration line. The observation that predation avoidance was incomplete in both lines—neither absolute success nor total failure—introduced significant ambiguity in the interpretation of these outcomes.

5.9 When Results Remain Ambiguous, it is Prudent to Return to First Principles

Assuming that death feigning represents an optimal antipredator strategy against jumping spiders, all individuals in the long-duration line would be expected to avoid predation. Yet, a subset of individuals in the short-duration line that do not perform death feigning also survived. While statistically significant differences between the lines were evident, the incomplete protection provided by death feigning remained puzzling. The question of why this behavior does not afford absolute survival continued to generate conceptual ambiguity.

When ambiguities persist, it is a fundamental tenet of research to return to first principles and perform direct experimentation. Accordingly, I isolated myself in a constant-temperature insect-rearing chamber and conducted repeated observations of *Tribolium castaneum* cohabiting with jumping spiders in a single Petri dish, meticulously recording the complete predatory interactions. During this iterative process, I made a significant and revealing observation.

As visually guided predators, jumping spiders are highly responsive to movement in potential prey, including *Tribolium castaneum*. During observation sessions, inadvertent human presence occurred when a student entered the controlled-temperature chamber and passed near the Petri dish. In these instances, the spider reoriented its attention from the beetle to the transient visual stimulus represented by the student's shadow, resulting in a temporary cessation of predatory focus on the beetle.

That's it!!!

Immediately, the observation prompted the generation of multiple predictive scenarios. I re-examined the ecological context in which *Tribolium castaneum* encounters jumping spiders as predator and prey. It became evident that a direct, one-on-one interaction between these species in natural settings is highly improbable, raising questions about the ecological relevance of experimental observations conducted under such simplified conditions. *Tribolium castaneum* does not occur as solitary individuals within rice bran; rather, they reproduce within rice or wheat flour, forming dense aggregations. The predation experiments reported in 2004 were conducted under one-on-one predator–prey conditions. It thus became evident that these experimental setups constitute highly artificial scenarios, rarely, if ever, encountered in the beetle's natural habitat. The oversight of this fundamental ecological context was striking.

We subsequently performed a modified experimental setup, introducing one adult from the short-duration line and one from the long-duration line into a single Petri dish, followed by the introduction of a single Adanson's jumping spider, thereby establishing a two-prey, one-predator configuration. Consistent with previous 2004 findings, the long-duration beetle exhibited death-feigning behavior in response to the spider's attack, remaining immobile. The spider initially maintained its focus on the stationary beetle; over time, the actively moving short-duration beetle entered the vicinity of the predator, approaching the immobile long-duration individual (Fig. 5.2).

As a result, the spider's focus shifted from the motionless prey to the actively moving short-duration beetle. The predator reoriented its body, with anterior visual organs tracking the moving short-duration individual rather than the stationary long-duration beetle. Upon the approach of the short-duration beetle, the spider initiated an attack. Given that the short-duration beetle had been selectively bred to lack death-feigning behavior, it did not respond with immobility and continued vigorous movements. The spider subsequently identified the moving beetle as a viable prey item, executing repeated attacks and ultimately capturing and consuming it. In this context, the struggling short-duration beetle, unlike its immobile long-duration counterpart, became vulnerable prey despite its protective exoskeletal covering.

Fig. 5.2 When long-strain beetles are attacked by jumping spiders, they exhibit tonic immobility in response to the threat. The jumping spider, confronted with the immobile long-strain beetle, attempts to assess whether it is suitable prey. Eventually, a short-strain beetle, which is unable to feign death, approaches. The spider's attention shifts to the short-strain beetle, which it attacks. The attacked short-strain beetle struggles and does not exhibit tonic immobility. Observing this, the jumping spider becomes convinced that it is indeed prey, attacks again, and consumes it. Consequently, the long-strain beetle survives (Video 5.1)

During these interactions, the long-duration beetle, having exhibited death-feigning behavior, subsequently recovered and resumed locomotion, effectively escaping the predator without apparent consequence. This pattern was consistently replicated across repeated trials.

5.10 Selfish Prey: The Benefits of Group Living

To further investigate, 19 female Adanson's jumping spiders were collected from the field, and 19 predation trials were conducted, each comprising one long-duration beetle, one short-duration beetle, and one spider, thereby creating a three-specimen predator–prey system. Spiders, food-deprived for several days, invariably preyed upon the short-duration beetles in all 19 trials (100% predation), whereas all long-duration beetles survived, demonstrating the consistent protective effect of death-feigning behavior under these experimental conditions.

Tribolium castaneum exists in dense aggregations under natural conditions. The encounter between a jumping spider and these beetle aggregations is analogous to the children's game "Daruma-san ga koronda." Upon entering a storage environment, the spider identifies a beetle cluster and attacks an individual. The attacked beetle responds by exhibiting death-feigning behavior, remaining immobile. Simultaneously, other nearby beetles continue active movement. As a predator highly responsive to movement, the spider redirects its attention toward the mobile individuals. Accordingly, short-duration beetles, which are incapable of death feigning, are predated upon, whereas long-duration beetles survive by remaining immobile. This observation clarifies the previously unresolved question regarding the adaptive function of death feigning in a social context.

Revisiting flour mills and coin-operated rice polishing facilities with a systematic approach, it became evident that *Tribolium castaneum* coexists with other stored-product insects, including *Oryzaephilus surinamensis*, *Lasioderma serricorne*, various Ptinidae, and Lepidopteran larvae. This observation suggested a hypothesis: even in the absence of conspecifics performing death feigning, a *T. castaneum* individual might achieve predator avoidance by exhibiting immobility in the presence of heterospecific insects, potentially reducing detection by visually guided predators such as jumping spiders. This insight informed the design of the subsequent experimental protocol.

To test the effect of heterospecific presence, a long-duration death-feigning *Tribolium castaneum* was cohabited with a closely related species, *T. confusum*, in a Petri dish, and a field-collected Adanson's jumping spider was introduced. Across repeated trials, the predator predominantly attacked the actively moving *T. confusum*, sparing the immobile long-duration *T. castaneum*. Conversely, in trials pairing a short-duration *T. castaneum* incapable of death feigning with *T. confusum*, predation was distributed approximately equally between the two species. From nearly 20 trials, these findings demonstrate that the presence of moving heterospecifics, rather than conspecifics alone, can divert predator attention and enhance the survival probability of death-feigning individuals.

These results indicate that, within aggregations—including heterospecific individuals—death-feigning *Tribolium castaneum* can survive only when surrounded by nearby, actively moving conspecific or heterospecific insects. In other words, death-feigning individuals can be considered "selfishly surviving at the expense of their neighbors." This finding suggests that the adaptive efficacy of death-feigning behavior is maximized in social or group-living contexts.

These results were synthesized into a manuscript entitled "Selfish Prey that Suddenly Become Immobile Survive at the Expense of Their Neighbors," published in 2009 in the *Proceedings of the Royal Society B*. The handling editor, Professor Graham Luxton, expressed considerable appreciation for the study. The conceptual framing of "selfish death feigning" garnered attention from the global scientific community, and a concise summary of the work was subsequently published on the online platform of the American general science journal *Science*, disseminating the findings to an international audience of researchers.

5.11 The Advantage of Being in the Minority

Although Adanson's jumping spiders can occasionally be found in stored-product facilities, field surveys indicated their presence at only a minority of sites (~3–4 of 50), generally with only one or two individuals observed per location. Additionally, the apparent absence of spiders at certain sites cannot be taken as definitive, given the possibility of undetected individuals. Therefore, direct comparisons of *Tribolium castaneum* behavior between spider-present and spider-absent environments are likely infeasible in natural contexts.

During this period, the predatory bug *Amphibolus venator*, previously reported from Okinawa, was identified as a potential new predator of *Tribolium castaneum*. To assess its distribution, surveys were conducted in over ten rice storage facilities on Okinawa Island and Ishigaki Island, following direct contact with facility managers. *T. castaneum* was commonly observed, whereas *A. venator* was rarely detected. Subsequent interviews indicated that the presence of *A. venator* within stored-product warehouses is highly localized.

Consequently, I decided to directly contact a rice storage company on Okinawa Island, where reports suggested that *A. venator* had been previously collected. Typically, personnel at food-processing companies are reluctant to permit insect surveys within their facilities. However, in this case, local connections facilitated access: a graduate of the Faculty of Agriculture at the University of the Ryukyus, where I had studied, managed the warehouse, and was a mutual acquaintance of a friend. Through the facilitation of a personal introduction, permission was granted to access the warehouse. Under the supervision of company staff, I surveyed an unused section of the facility and observed a high density of both *T. castaneum* and predatory bugs, *A. venator*, moving actively across the rice flour substrate, thereby confirming the sympatric occurrence of the prey and its putative predator.

With the company's permission, I collected both *A. venator* and *T. castaneum* from the warehouse. Subsequently, I reared *T. castaneum* individuals collected from areas where they co-occurred with *C. grandis* and from areas without co-occurrence, and compared the death-feigning behavior between these two groups. Predator experience markedly influenced the death-feigning response of *T. castaneum*. When subjected to a standardized stimulus, individuals originating from populations cohabiting with *Amphibolus venator* exhibited significantly prolonged immobility relative to individuals from populations lacking prior predator exposure. Quantitatively, beetles from predator-associated populations demonstrated an average death-feigning duration of ~400 seconds, whereas beetles from predator-free populations displayed an average duration of merely ~5 seconds.

The findings indicate that *T. castaneum* subjected to frequent predation in natural environments are more likely to exhibit death-feigning behavior, implying that this trait is subject to positive selection. Consequently, generational exposure to natural selection may shift the frequency of death-feigning individuals within populations. In this sense, the results offer indirect evidence supporting the evolutionary adaptation of death feigning. A subsequent follow-up to this observation is described below.

5.12 Death-Feigning Response to Jumping Spiders and Freezing Response to Predatory Stink Bugs

Both mammals and insects exhibit a sequence of antipredator responses when threatened by predators. For instance, small mammals and insects initially employ a freezing response upon detection of a nearby predator. If the predator continues to approach, the prey may shift to active defense or attempt to flee. When escape is no longer possible, some species adopt a characteristic death-feigning posture, and in certain cases, such as snakes or opossums, death feigning is accompanied by the release of noxious odors. It is widely recognized among researchers that prey organisms typically display a sequential progression of antipredator behaviors, with initial freezing responses occurring upon predator detection, and death-feigning manifesting at later stages. Consequently, death feigning is often characterized as a "last-resort" behavior in the literature. This framework is endorsed by Ruxton et al., and by an emeritus professor at Okayama University who provided instruction on the physiological basis of death feigning. In his 2021 monograph, Death-Feigning in Insects: Mechanism and Function of Tonic Immobility, this sequence is outlined in the Introduction (Sakai 2021). Prior to my investigations, I had not anticipated challenging the prevailing assumption that individuals capable of death feigning invariably employ it under life-threatening conditions, such that no individuals abstain from this last-resort strategy. Following extensive predation trials using Adanson's jumping spider as a model predator for *Tribolium castaneum*, we subsequently employed the laboratory-easy-to-maintain predatory bug for undergraduate thesis experiments. In 2014, we devised an experimental evolution framework to test whether multigenerational cohabitation with the predatory bug would alter the frequency of death feigning in *T. castaneum*. Through careful planning with students, we established multiple experimental populations co-reared with *Amphibolus venator* for approximately eight generations, alongside control populations maintained without predator exposure. After several generations, we assessed and compared the evolution of antipredator strategies, focusing on death feigning, between predator-exposed and predator-free populations. However, even after maintaining *T. castaneum* in cohabitation with *A. venator* for multiple generations to experimentally simulate evolution, no evolutionary response—in terms of increased frequency of death feigning—was observed. In the undergraduate thesis, it was reported that co-rearing with the predator *A. venator* did not induce measurable changes in the antipredator behaviors of *T. castaneum*, including death feigning. The discussion suggested that the absence of a response might be due to the limited number of generations used in the experimental evolution. Nevertheless, the complete lack of observable change remained puzzling. At one point, we decided to closely observe the behavioral responses of *T. castaneum* when attacked by *A. venator*. The results were both disappointing and surprising. In early spring 2021, we found that *T. castaneum* exhibited no death-feigning behavior in response to attacks by *A. venator*. While it was expected that the short-strain beetles would not perform death feigning, it was remarkable that even the long-strain individuals failed to do so when the predator was the predatory bug. How, then, did the long-strain *T. castaneum* avoid predation by *A. venator*? When the predatory bug approached, the beetles simply remained immobile, patiently waiting for the predator to pass by. Faced with this unexpected result, we exposed 20 long-strain individuals, one at a time, to different *A. venato*r predators. In each case, the beetles exhibited a freezing response, occasionally moving slowly, but never assuming the characteristic death-feigning posture with their ventral side exposed. Furthermore, in the long-strain *Tribolium freemani*, which was established using the same selective breeding as for *T. castaneum*, in 1–2 out of 20 trials, the *A. venator* predator appeared to make contact with the beetle and suddenly exhibited convulsive reactions. This was likely due to the release of the defensive chemical, methyl-1,4-benzoquinone, by the beetle. Notably, such a response was rarely observed when the predator was Adanson's jumping spider.

I hypothesize that the key to this phenomenon lies in the differences between the stimuli experienced by *T. castaneum* when attacked by Adanson's jumping spider versus when contacted by *A. venator* attempting to feed. Regardless of the number of observations, even though *A. venator* made body contact with the beetles and, in some cases, successfully captured them; the long-strain beetles never exhibited death-feigning behavior in response to the stimuli provided by *A. venator*. *Tribolium castaneum*, measuring less than 5 mm in body length, exhibits predator-specific avoidance strategies. When attacked by the active-hunting predator, Adanson's jumping spider, the beetles perform death feigning, whereas when confronted by the ambush predator, *A. venator*, they exhibit freezing behavior. This finding challenges the previous consensus in the field of death-feigning research, which assumed a fixed sequence in which prey first freezes upon predator approach and resorts to death feigning only as a last resort. Our results demonstrate that *T. castaneum* flexibly adjusts its antipredator strategy depending on the type of predator encountered. This work was recently published in the journal *Ecology and Evolution* in early 2022 (Asakura et al. 2022). The media has sometimes described *Tribolium*

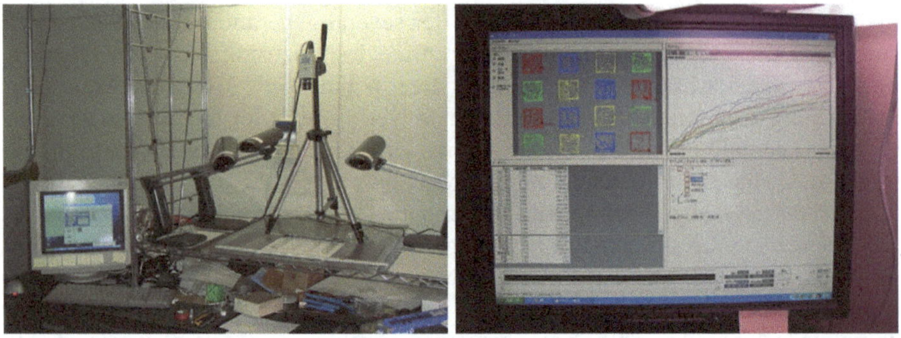

Fig. 5.3 *Tribolium castaneum* exhibiting freezing behavior in the presence of the predatory assassin bug, *Amphibolus venator*

castaneum as a "dramatic" insect, which generated considerable public interest. A new question arising from our work is why the beetles do not exhibit death-feigning behavior in response to stimuli from *A. venator*. In addressing such puzzling observations, it remains crucial to return to first principles and conduct careful, detailed behavioral observations. I believe that this approach represents the most reliable path to uncovering true biological mechanisms. In an era where genomic analyses can reveal much about the underlying machinery of life, returning to direct observation of behavior has become even more important (Fig. 5.3).

Recently, experimental and control groups were established in which *Amphibolus venator* and *Tribolium castaneum* were either cohabitated or kept separate, and experimental evolution was conducted over five generations. The results revealed a surprising outcome: in the cohabitated group, *Tribolium castaneum* exhibited a significantly shorter duration of tonic immobility compared to the non-cohabitated control group (Miura et al. 2025).

5.13 Indirect Reactions (Trade-Off)

Is increased survival the only benefit? The establishment of experimentally bred lines that exhibit or lack death-feigning behavior provides a valuable framework for addressing numerous unresolved questions. While individuals that feign death are less likely to be predated upon, it remains unclear whether this behavior confers additional fitness advantages. For instance, how do death-feigning and non-death-feigning strains differ in terms of developmental rate, somatic growth, or reproductive output? Do individuals exhibiting thanatosis demonstrate enhanced performance in aspects of breeding success or offspring viability?

As an initial approach, we compared the adult lifespan of strains exhibiting short and long durations of death-feigning behavior. Although we first attempted this comparison using *Tribolium castaneum*, the species' exceptionally long adult lifespan—often exceeding 1 year—rendered it impractical for such an investigation.

Consequently, we turned to the adzuki bean beetle (*Callosobruchus chinensis*), a more suitable model organism. In this species, artificial selection has successfully produced strains with genetically determined differences in death-feigning duration (Ohno and Miyatake 2007), allowing us to assess whether this behavioral trait is associated with variation in lifespan.

A comparison of life-history traits, including developmental time and adult longevity, between short and long death-feigning strains of the adzuki bean beetle (*Callosobruchus chinensis*) revealed notable differences. Individuals from the short-strain lineage, which exhibit minimal or no death-feigning behavior, were typically highly active and mobile. In contrast, individuals from the long-strain lineage, which displayed prolonged death feigning, were generally less mobile (Nakayama and Miyatake 2009a). This reduced mobility suggests a lower baseline energy expenditure, whereas high mobility in short-strain individuals likely reflects increased energy consumption through frequent locomotion.

Moreover, individuals from the long death-feigning strain exhibited several advantageous life-history traits compared to those from the short strain. The long-strain beetles developed more rapidly, requiring less time to transition from egg to adult. After reaching adulthood, they also demonstrated greater longevity than their short-strain counterparts. In addition, the long strain showed a higher pupal-to-adult emergence rate. Although the total number of eggs laid did not differ significantly between strains, females of the long strain produced larger eggs. Since larger egg size is generally associated with higher offspring survival, this trait may contribute to greater overall fitness. Taken together, these findings indicate that the long-death-feigning strain possesses a consistent advantage across multiple life-history parameters (Nakayama and Miyatake 2009a).

It is intuitively plausible that death-feigning behavior requires less energy than active escape responses such as running. By remaining motionless until a threat subsides, individuals can minimize energy expenditure, engaging in movement only when sensing immediate danger or upon direct contact with a predator. This energy-conserving strategy may partially explain the life-history advantages observed in the long-death-feigning strain. These individuals are not only less likely to be consumed by predators but also exhibit faster developmental rates, increased adult longevity, and higher pupal-to-adult emergence success. Additionally, females from the long strain produce larger eggs, which are generally associated with enhanced offspring survival. Collectively, these traits suggest that prolonged death-feigning confers substantial adaptive advantages beyond immediate predator avoidance.

At first glance, individuals that exhibit prolonged death-feigning behavior appear to possess comprehensive life-history advantages. If this were universally true, one might expect all insects to evolve such a strategy. However, biological systems are rarely so straightforward. In nature, it is highly unlikely for any single behavioral or physiological strategy to be optimal across all fitness components. Trade-offs are an inherent feature of evolutionary processes.

Indeed, if prolonged death feigning were advantageous in every respect, we would expect all individuals within a species to evolve this trait. Yet, field-collected populations of adzuki bean beetles and *Tribolium* species consistently exhibit considerable variation in death-feigning duration (Konishi et al. 2020). As noted earlier in this chapter, organisms must allocate finite resources among competing survival and reproductive demands. This implies that, despite its apparent benefits, prolonged death feigning likely incurs costs in other domains—costs that maintain the observed behavioral diversity through evolutionary trade-offs.

5.14 Never Met Anyone of the Opposite Sex...

Ultimately, a significant cost associated with death-feigning behavior was identified. Individuals that exhibit prolonged tonic immobility typically show reduced locomotor activity (Miyatake et al. 2008). While reduced movement may decrease the likelihood of detection by predators—thus enhancing survival—it also limits opportunities to encounter potential mates. In insects, which lack social communication technologies or long-distance signaling systems, remaining immobile is analogous to remaining isolated; mating opportunities are severely diminished.

Regardless of how long an individual survives, sexually reproducing organisms must engage in mating and produce offspring for their genetic material to persist. From an evolutionary perspective, fitness is a function of both survival and reproductive success. If an individual fails to reproduce—despite high survival—its overall fitness remains effectively zero.

The following is a description of the experimental methodology used to identify the reproductive disadvantages associated with prolonged death-feigning behavior. The experiment was conducted using laboratory-bred *Tribolium castaneum* strains that had undergone over 35 generations of artificial selection for either long or short durations of tonic immobility.

Sex differentiation in *T. castaneum* is possible at the pupal stage based on the morphology of the terminal abdominal segments, but it becomes difficult once the beetles reach adulthood. Therefore, 1 day prior to the experiment, adult males and females were sex-marked by applying colored paint dots to the dorsal surface of the thorax using fine-tipped markers.

In each trial, five virgin females and one virgin male—either from the long- or short-death-feigning strain—were placed together in a 9-cm-diameter Petri dish. The number of successful matings (i.e., the number of females with which the male copulated) was recorded. Observations were conducted by a postgraduate student in the laboratory and myself.

During a 15-minute observation period, males from the short-death-feigning strain successfully copulated with an average of 3.5 females, whereas males from the long-death-feigning strain mated with approximately 2 females on average (Nakayama and Miyatake 2009b). While this may seem like a modest difference—only 1.5 females—the implications for reproductive output are substantial. Even small differences in mating frequency, when extrapolated across a male's lifetime, can result in large disparities in total offspring production.

In a separate experiment, adults of both strains were exposed to the jumping spider *Hasarius adansoni* under cohabitation conditions. In this predator-exposure assay, long-strain individuals survived significantly longer than those from the short strain. These results highlight a clear evolutionary trade-off: individuals from the long-death-feigning strain enjoy greater predator avoidance but suffer reduced mating success due to decreased activity and fewer encounters with potential mates. This set of findings provides direct empirical evidence of the cost–benefit balance that maintains behavioral polymorphism in natural populations.

These findings were published in the journal *Biology Letters*, published by the Royal Society, under the title "Genetic trade-offs between predation avoidance and mating ability: the cost of tonic immobility" (Nakayama and Miyatake 2009b) in the summer of 2009. The results illustrated a striking example of evolutionary compromise. As the saying goes, when one aspect is optimized, another may be constrained. In this case, the prolonged death-feigning behavior that effectively reduces predation risk simultaneously imposes a reproductive cost by limiting encounters with potential mates. Thus, even this seemingly adaptive antipredator strategy is subject to a fundamental trade-off that likely contributes to the maintenance of behavioral variation in natural populations.

5.15 Indirect Responses Physiology/Mechanisms

They were also vulnerable to stress… During the publication process of the aforementioned series of studies, a professor specializing in insect physiology at Saga University expressed strong interest in the phenotypic characteristics of the long- and short-duration strains. Around 2010, he proposed conducting a comparative assessment of their stress tolerance. I readily agreed and subsequently sent both strains—one exhibiting prolonged death feigning and the other exhibiting brief death feigning—to his laboratory for further analysis.

The initial focus of the professor's investigation was a comparison of survival rates between the short- and long-duration strains under two types of stress: mechanical vibration and low temperature. Adult beetles from both strains were placed in Petri dishes and exposed to intense mechanical stimulation at a frequency of 2000 strokes per minute. After 4 hours of continuous vibration, approximately 50% of the long-strain individuals—characterized by prolonged death feigning—had died, and mortality increased to approximately 80% after 8 hours. In contrast, about 70% of the short-strain individuals survived even after 8 hours of vibration exposure (Kiyotake et al. 2014).

Low-temperature tolerance was also assessed. When adult beetles were maintained at 4 °C, a temperature comparable to standard refrigeration, approximately 60% of the long-strain individuals died within 4 days. In contrast, it took over 10 days for 60% mortality to occur in the short-strain individuals. Additionally, thermal stress tolerance was evaluated by exposing individuals to high temperatures. The long-strain beetles, characterized by prolonged death-feigning behavior, also exhibited increased susceptibility to heat stress at 45 °C. These findings indicate that the acquisition of traits associated with extended death-feigning behavior was consistently accompanied by reduced tolerance to multiple environmental stressors (Kiyotake et al. 2014).

The research team at Saga University subsequently conducted a comparative analysis of the expression levels of several stress-related genes between the long- and short-duration strains. They found that expression of the *catalase* gene—which promotes the breakdown of hydrogen peroxide generated within the organism—was significantly higher in the short-strain individuals. Additionally, expression of the gene encoding *GDP (Guanosine diphosphate)*, which is implicated in the regulation of insect larval development, was also elevated in the short strain compared to the long strain (Kiyotake et al. 2014).

In natural populations of *Tribolium castaneum*, such as those inhabiting grain storage warehouses, the majority of individuals exhibit death-feigning durations similar to those of the short-duration laboratory strain. These individuals rarely enter tonic immobility, or if they do, recover within a few seconds (Kiyotake et al. 2014; Konishi et al. 2020). Although a small number of individuals display prolonged death-feigning behavior resembling that of the long-duration strain, they constitute a minority within the population. This distribution suggests that individuals exhibiting prolonged tonic immobility may be physiologically disadvantaged in terms of stress resistance and reproductive fitness in environments where predation pressure is low or absent. Elevated expression of stress-related genes such as *catalase*, along with genes associated with growth suppression, may enhance stress tolerance under natural conditions, thereby contributing to the predominance of short-duration death-feigning phenotypes in the field.

In other words, the long-duration strain appears to possess a physiologically weaker constitution under various stress conditions. In addition to their reproductive disadvantage, individuals that exhibit prolonged death-feigning behavior are also more susceptible to physiological stress. This dual vulnerability may help explain why such long-death-feigning phenotypes are relatively rare in natural populations of *T. castaneum* (Kiyotake et al. 2014).

5.16 Heaven Gave Two Gifts

However, findings reported in 2020 suggest that, in some cases, individuals with a reproductive advantage can also effectively avoid predation by exhibiting death-feigning behavior (Matsumura et al. 2020). The broad-horned flour beetle (*Gnatocerus cornutus* Fabricius), a pest species that infests flour and other stored food products, exhibits sexual dimorphism: only males possess large, well-developed mandibles, which they use in intrasexual combat to secure mating opportunities (Okada and Miyatake 2006). When threatened, *G. cornutus* also displays tonic immobility, retracting its legs and adopting a death-like posture similar to that of *Tribolium castaneum*. The duration of death-feigning behavior varies considerably among individuals, from complete absence of the response to immobility lasting up to 200 seconds. On average, individuals exhibited tonic immobility for approximately 60 seconds, with the modal duration being around 40 seconds.

The duration of death-feigning behavior in male individuals following stimulation was measured; however, no significant correlation was found between body size and the duration of tonic immobility.Surprisingly, a positive correlation was observed between the frequency of death feigning and mandible size: males with larger mandibles were more likely to exhibit tonic immobility in response to external stimuli.In contrast, no association was found between thorax width—used as an index of overall body size—and the frequency of death-feigning behavior (Matsumura et al. 2020).

These findings suggest that while the *duration* of death feigning is independent of body size, the *frequency* of this behavior is influenced by specific morphological traits, such as mandible size.

This experiment demonstrated that individuals exhibiting greater success in male-male combat also showed a higher propensity for death-feigning behavior. These findings challenge the commonly assumed trade-off between reproductive advantage and predator avoidance, suggesting that both traits can co-occur in certain individuals. The results were published as a brief report in *Biology Letters*, a journal of the Royal Society (Matsumura et al. 2020), in the spring of 2021.

These findings may lead readers to question whether the biological principle of trade-offs applies to *Gnatocerus cornutus*, given that males with larger body size not only have a mating advantage but also appear to be more adept at avoiding predation through death-feigning behavior. To address this potential confusion, several additional points require clarification.

Males of *Gnatocerus cornutus* engage in aggressive interactions using their well-developed mandibles immediately upon encountering one another, even in the absence of females. The victor of such contests subsequently mates with any females present within his territory. In contrast, defeated males exhibit prolonged avoidance behavior, refraining from further confrontations for approximately 4 days. On the fifth day, however, they resume fighting with rival males as if the previous defeat had not occurred, indicating that the memory of social defeat persists for only about 4 days (Okada and Miyatake 2010).

Interestingly, when a defeated male encounters a female during this avoidance phase, he still engages in courtship and copulation. Moreover, such males have been shown to increase their ejaculate investment by releasing a larger quantity of sperm (Yamane et al. 2010). These findings suggest that male *G. cornutus* exhibit a sexually selected mating system that incorporates territorial behavior and plastic reproductive strategies in response to social context.

Larger males of *Gnatocerus cornutus* are dominant in male–male competition but are less effective in courtship interactions with females (Harano et al. 2010). This suggests a trade-off in males of this species: while they excel at avoiding natural enemies—possibly due to their greater propensity for death-feigning behavior—they perform poorly in female courtship.

As with many organisms constrained by finite biological resources, it remains unlikely that individuals can optimize all fitness-related traits simultaneously. Further research is needed to clarify how environmental conditions in the field—particularly those related to predation pressure—shape the balance between reproductive success and survival strategies, including tonic immobility. In the following section, I present another example in which a behavioral dichotomy was clearly observed: individuals either remained in motion without feigning death or switched to immobility and feigned death. The discovery of this pattern was initially prompted by a scientific meeting in 2003, which I still vividly remember.

5.17 Beetles That Are Good Fliers Tend Not to Feign Death

During a poster presentation at the annual meeting of the Japanese Society for Ethology, I received an insightful comment from an attendee regarding insect antipredator behavior. They remarked, "When I collect insects in the field, those with strong flight ability tend to escape by flying away. In contrast, poor fliers often drop to the ground and remain motionless." This observation resonated with my own field experiences. For example, when attempting to collect beetles resting on leaves, they frequently drop suddenly to the ground or into the grass, disappearing from view. Many of these insects remain immobile after falling, suggesting the use of death-feigning behavior as an escape strategy.

In contrast, insects such as butterflies and dragonflies almost invariably respond to capture attempts by rapidly flying away—making them far more difficult to catch. These anecdotal observations suggest that insects with limited flight ability may rely more heavily on thanatosis as an alternative antipredator strategy.

Flying and death feigning represent two fundamentally distinct antipredator strategies: the former involves active movement, while the latter is characterized by complete immobility. Since death feigning cannot be performed during flight, an insect under predation threat faces a binary choice—to flee by flying or to adopt tonic immobility.

This raises an intriguing question: does this trade-off operate not only at the interspecific level but also within species? Specifically, might insects artificially selected for prolonged death-feigning behavior exhibit reduced flight ability, whereas those selected for enhanced flight performance show diminished capacity for tonic immobility?

With this conceptual framework in mind, I immediately considered investigating this behavioral dichotomy within a breeding line of *Tribolium castaneum*. However, *T. castaneum* is generally a poor flyer under standard laboratory conditions, although flight ability improves markedly at temperatures above 28 °C.

Upon my appointment at Okayama University, I expanded my research to include the adzuki bean weevil (*Callosobruchus chinensis*), as detailed in Chap. 2. While my initial interest in this species stemmed from its intriguing oviposition and mating behaviors, I found that the adzuki bean beetle is not only easy to rear in the laboratory but also exhibits surprisingly strong flight capabilities.

The experimental methods and supporting data for the following statements are detailed in Ohno and Miyatake (2007). The adzuki bean beetle (*Callosobruchus chinensis*) was identified as an ideal model organism to investigate the genetic dichotomy between flight ability and death-feigning behavior, as it exhibits both traits and can be readily collected from natural populations. Shortly after selecting this species, I initiated discussions with a postgraduate student in the laboratory to design the experiment. One approach to assess whether flight and death feigning are genetically distinct traits involves controlled breeding experiments, a methodology previously applied to *Tribolium castaneum*. Consequently, the graduate student developed a standardized assay to evaluate flight performance in the adzuki bean beetle. The apparatus consisted of a transparent acrylic cube measuring 50 cm on each side, with a small hole (7 mm diameter) drilled at the center of the top panel through which adult beetles were released. The bottom surface of the cube was marked with five concentric circles spaced at 5-cm intervals to record the horizontal distance flown after release.

Points were assigned based on the flight performance of each dropped adult beetle as follows: zero points were given if the beetle fell directly—without adopting a landing posture—into the central circle measuring 10 cm in diameter. One point was awarded if the beetle attempted to fly and landed within the 10 cm circle before stopping. Two points were given if the beetle landed within the 20-cm-diameter circle, three points within the 30 cm circle, four points within the 40 cm circle, and five points within the 50 cm circle. The highest score of six points was assigned if the beetle flew beyond the 50 cm circle and either landed near the acrylic cube's wall or on its surface. This scoring system provides a quantitative measure of the flight ability of the adzuki bean beetle.

A total of 100 individuals, consisting of 50 males and 50 females, were assessed for their flight ability using the scoring system described above. The seven males and seven females with the highest scores were selected as breeders and allowed to mate freely to produce the next generation. Conversely, the seven males and seven females with the lowest scores were also selected and allowed to mate freely to establish a lineage of poor fliers. This selective breeding protocol was repeated for 13 generations to establish genetically distinct populations characterized by strong or weak flight capabilities. In each generation, in addition to measuring flight performance, the duration of death-feigning behavior was also recorded using a stopwatch.

The populations selectively bred for high flight ability exhibited progressively higher flight scores over successive generations, while the duration of death-feigning behavior correspondingly decreased with each generation. Conversely, populations selected for low flight scores became increasingly flightless, and by the seventh generation, these beetles exhibited complete loss of flight ability. This flightless group simultaneously displayed prolonged durations of death-feigning behavior. These trends were observed consistently in both males and females. Furthermore, the breeding experiments were independently replicated twice, yielding consistent results across all trials.

It is clear that the well-flying populations became higher and higher flight-score populations over the generations, but in contrast, the duration of death feigning became shorter and shorter each generation. Conversely, populations bred in the direction of low flight scores became increasingly flightless, and by the end of seven generations had become a population of insects that did not fly at all. The insects, which then became a flightless group, began to exhibit a long duration of death feigning. This result was the same for both males and females, and further breeding experiments were conducted twice independently, with the same results both times.

The populations selectively bred for high flight ability exhibited progressively increasing flight scores over successive generations, while the duration of death-feigning behavior decreased correspondingly. Conversely, populations selected for low flight scores became increasingly flightless, and by the seventh generation, they comprised individuals that no longer exhibited flight. These flightless beetles concurrently displayed prolonged durations of death-feigning behavior. This pattern was consistent across both males and females. Moreover, the breeding experiments were independently replicated twice, with consistent results obtained in both trials.

It is now evident that selection for increased flight ability results in a correlated, inverse response in the duration of death-feigning behavior. To conclusively determine whether flight ability and death feigning represent genetically antagonistic traits, a complementary breeding experiment is necessary. Specifically, selecting for variation in death-feigning duration and assessing the correlated response in flight ability will clarify whether these traits exhibit a negative genetic trade-off.

Accordingly, the graduate students initiated selective breeding for both longer and shorter durations of death-feigning behavior in the adzuki bean beetle. This selection experiment was successfully conducted for eight generations, with flight scores measured concurrently throughout the breeding process.

The results demonstrated that populations selected for prolonged death-feigning behavior became completely flightless within only four generations. Conversely, in lines selected for reduced death-feigning duration, males that initially had an average flight score of approximately 2 increased to over 4 after eight generations, while females with an initial average flight score of around 1.5 increased to over 3 after eight generations (Ohno and Miyatake 2007).

5.18 The Trade-Off Worked in the Field Populations.

The results of the two breeding experiments described above confirm the existence of a negative genetic correlation between flight ability and death-feigning behavior. If this genetic trade-off operates under natural conditions in *Callosobruchus chinensis* populations, then a similar inverse relationship between these two traits should be observable within wild-caught beetle populations.

In autumn, we collected *Callosobruchus chinensis* from multiple adzuki bean fields across Okayama Prefecture. Additionally, I collaborated with colleagues in insect research, requesting that they send adzuki bean beetles collected from adzuki and cowpea plants during the autumn season, as well as share laboratory strains. Through these efforts, we obtained 21 populations of adzuki bean beetles from various regions throughout Japan.Upon measuring both death-feigning duration and flight scores for each population, we observed a remarkably clear negative correlation between these two traits.

Males of this species generally exhibit superior flight ability compared to females. Consequently, the flight scores of well-flying male groups exceed 4, whereas even the best-performing female groups typically reach only around 2.5. Despite this sexual dimorphism in flight performance, a strong negative correlation between death-feigning duration and flight score was observed in both sexes. In other words, populations that exhibited greater flight ability tended to display shorter durations of death-feigning behavior, while those with lower flight scores exhibited prolonged death feigning.

Although the ecological factors driving the formation of high- and low-flight groups in the field remain unclear, variations in predator communities are likely contributors. Importantly, this study provides compelling evidence that, at the population level within a single species, individuals with strong flight capability are less capable of death feigning, whereas flightless individuals compensate by exhibiting prolonged thanatosis.

The results of this series of studies were succinctly summarized under the title "Drop or Fly? Negative Genetic Correlation between Death-Feigning Intensity and Flight Ability as Alternative Anti-Predator Strategies," published in *Biology Letters* in November 2007 (Ohno and Miyatake 2007).

5.19 What Happens Within the Organism? Mechanistic Insights

Let us return to the early summer of 2003, 2 years after we initiated selective breeding for death-feigning duration. One of the graduate students involved in the breeding experiments came to my office and exclaimed, "Sensei, the larvae are freezing like branches" (Sensei means Teacher in Japanese).

I was struck by the vividness of this observation—it was truly remarkable…

The larvae then began to mimic death by stiffening their bodies, resembling branch-mimicking cankerworms—rigid, stick-like, and remaining erect and motionless. At that moment, I realized something important: we had been selecting adult beetles based on their death-feigning response to stimulation, but we had never directly stimulated or selected for larval behavior. This unexpected observation suggested a possible correlated response in the larvae.

At that moment, a question arose in my mind. As beetles are holometabolous insects undergoing complete metamorphosis with a pupal stage between larva and adult, does the death-feigning behavior persist unchanged through metamorphosis, or is the larval stiffening a distinct phenomenon? If the behavior is consistent across stages, an alternative explanation might be that a physiological substance present within the body of *T. castaneum* influences body rigidity throughout metamorphosis. This possibility sparked my interest in identifying the substance or underlying mechanism that regulates the duration of death-feigning behavior in *T. castaneum*.

With limited knowledge of physiology, I sought guidance from a professor specializing in insect physiology at the Faculty of Science, Okayama University. I consulted him regarding the substances that regulate insect activity. This professor had conducted neurophysiological research on death feigning and mating behaviors, primarily using crickets and grasshoppers as model organisms. He is recognized as a pioneer in the study of death-feigning behavior in Japan.

The professor, known among his students as a hermit due to his distinctive white beard and reclusive demeanor, received my questions with surprising warmth. I nervously asked which substances regulate insect activity. He explained in a friendly manner that biogenic amines such as octopamine, dopamine, and tyramine promote insect activity, whereas serotonin generally induces inactivity. These neurotransmitters play a crucial role in controlling insect movement. This encounter took place in the autumn of 2003. The professor's name is Dr. Masaki Sakai, who later authored a pioneering and highly regarded book on death-feigning behavior (Sakai et al. 2021).

In the following spring of 2004, I attended a meeting of entomologists held in Kyoto, where several presentations focused on insect dopamine were delivered. One presentation particularly impressed me; the researcher measured the concentrations of biogenic amines in the brains of bees. This work led me to consider that if it is possible to quantify biogenic amines in the relatively large brains of bees, then it might also be feasible to measure these substances in the much smaller brains of *Tribolium castaneum* beetles.

After the lecture, I approached the researcher and asked, "Would it be possible to measure the biogenic amines expressed in the brains of the death-feigning and non-death-feigning strains of *T. castaneum*?" This was a spontaneous question to someone I had never met before. The researcher replied, "Biogenic amines are expressed in the insect brain. How large is *T. castaneum*?" When I told him its size, he responded, "It's so small that measuring biogenic amines would be quite challenging and risky."

Understandably, my question came out of the blue from someone the researcher did not know well, and measuring biogenic amines in such a small insect was indeed a challenging proposition. However, the researcher suggested an alternative approach:"It is risky to immediately attempt measuring a substance without prior knowledge. But could you at least try injecting some of the biogenic amines into the beetle's body? That way, we might gain insight into the substance's effects and start forming hypotheses."

Upon returning from the conference, I shared this story with a graduate student in my laboratory, who immediately became interested. They proceeded to design an experiment to investigate the effects of biogenic amine administration on adult beetles from the long-duration death-feigning strain. The amines selected for testing were octopamine, dopamine, tyramine, and serotonin, as previously discussed. These compounds were obtained in powder form, dissolved in distilled water, and then injected into the insects. The injections were performed under a stereomicroscope using a glass capillary tube with an extremely fine, tapered tip, targeting the intersegmental membrane between abdominal segments.

This task proved challenging for me, especially given my already advanced presbyopia at the time, but the graduate students performed the injections skillfully and efficiently. When the duration of death-feigning behavior was measured in adults injected with each biogenic amine, as well as distilled water as a control, the duration for serotonin-injected and control beetles remained unchanged. In contrast, adults injected with octopamine, tyramine, and dopamine exhibited significantly shorter death-feigning durations. Subsequently, different concentrations of these biogenic amines were tested. Notably, increasing the concentration of dopamine from 1% to 5% and 10% resulted in a dose-dependent reduction in death-feigning duration (Nishi et al. 2010).

5.20 What Happens in the Brain?

It was hypothesized that dopamine plays a pivotal role in modulating the duration of thanatosis (death-feigning behavior). Following the completion of dopamine injection experiments, I shared the results with an insect physiologist I had previously consulted. Upon reviewing the data, he responded with a smile and suggested, "Let's proceed with a brain dissection." The researcher in question, Dr. Ken Sasaki—then an Associate Professor at the Kanazawa Institute of Technology and currently a Professor at Tamagawa University—subsequently received two genetically distinct strains of *Tribolium castaneum*, which I shipped to his laboratory in Kanazawa via refrigerated courier. The first strain (short-duration) comprised beetles that rarely exhibit thanatosis and respond to external stimuli with sustained activity, whereas the second strain (long-duration) consisted of individuals that display prolonged tonic immobility in response to even mild stimulation.

Tribolium castaneum beetles are approximately 4 mm in length. Dr. Ken Sasaki performed a dissection of the adult *T. castaneum* brain by opening the head capsule and carefully extracting the brain tissue. On October 30, 2005, I visited Dr. Sasaki's laboratory in Kanazawa, where he demonstrated the dissection procedure under a microscope. I was deeply impressed by the high level of technical expertise demonstrated by Dr. Sasaki.

Subsequent analyses employing liquid chromatography—an analytical technique used to quantify chemical compounds—were conducted using instrumentation provided by Dr. Sasaki. These analyses revealed distinct neurochemical differences between the two *T. castaneum* strains (see Miyatake et al. 2008). Specifically, the brains of short-strain beetles, which rarely exhibited thanatosis, contained significantly higher concentrations of dopamine, whereas the brains of long-strain beetles, which displayed prolonged death feigning, exhibited markedly lower dopamine levels. These findings were published in the international journal *Animal Behaviour* in 2008. Since that time, Dr. Sasaki and I have continued our collaborative research efforts.

Dr. Sasaki also measured the levels of other biogenic amines—octopamine, tyramine, and serotonin—in the brains of both death-feigning and non-death-feigning strains. The analysis revealed no significant differences in the concentrations of these substances between the two groups (Nishi et al. 2010). This result strongly suggested that dopamine was the primary neurochemical associated with variation in death-feigning behavior.Encouraged by how close we seemed to be to identifying the key mechanism, we began designing what we called a "no-go" experiment.

5.21 Behavioral Excitation in Beetles Following Caffeine Exposure

Dopamine, a neurotransmitter associated with arousal and excitement in the brain, is known to be released at elevated levels following the intake of certain substances—most notably caffeine. Caffeine, a well-known stimulant found in coffee and tea, enhances dopaminergic activity in both vertebrates and invertebrates. This led us to hypothesize that administering caffeine to *Tribolium castaneum* beetles from the long-duration death-feigning strain might increase their arousal level, thereby reducing their ability to remain motionless and sustain death-feigning behavior.

With this hypothesis in mind, my graduate students and I immediately began experiments using commercially available powdered caffeine. Although the concept seemed simple, it required some ingenuity to successfully administer caffeine to the beetles. Our initial attempt involved dissolving caffeine in water and applying it to a nonfat cotton ball; however, the beetles showed little interest—likely due to caffeine's bitterness. To overcome this, we mixed caffeine with sucrose and applied the solution to cotton. This time, the adult *T. castaneum* eagerly inserted their heads into the cotton and consumed the solution enthusiastically. This setup allowed us to compare the duration of death-feigning behavior in adults fed sucrose with caffeine versus those fed sucrose alone. The results clearly supported the stimulatory effect of caffeine on arousal from thanatosis. As anticipated, beetles from the long-duration strain—which typically remain motionless for extended periods—became active shortly after ingesting caffeine. Remarkably, both males and females from the long strain exhibited a tenfold reduction in death-feigning duration after caffeine intake. In other words, even insects become more active after a dose of caffeine—just like us after a cup of coffee.

Even insects become restless after consuming caffeine. As a side note, I recently supervised a student research project investigating how caffeine affects courtship behavior in male *T. castaneum*. Given that caffeine significantly reduces the duration of death feigning, it is not surprising that it may also influence other behavioral traits. These findings suggest that caffeine, through its action on the dopaminergic system, could have broad effects on behavioral expression in beetles beyond thanatosis.

Male *T. castaneum* beetles that were given caffeine exhibited significantly accelerated courtship behavior compared to non-caffeinated males. Specifically, the time required to initiate courtship and to mount a female after being introduced into a container was shorter in caffeinated males. In addition, the interval between first contact with the female and protrusion of the copulatory organs was also reduced, indicating that caffeine intake accelerates the entire sequence of courtship behaviors (Yuhao et al. 2021).

Furthermore, using a black body color mutant strain, an experiment was conducted in which a single female was mated with one caffeinated male and one non-caffeinated male to determine which male's sperm fertilized the eggs. As expected, caffeine had no effect on the fertilizing ability of the sperm. Nevertheless, the observation that caffeine influences insect mating behavior represents a novel and important finding. The results of this study were published in *Ethology* in the autumn of 2020 (Yuhao et al. 2021). While further research into the effects of caffeine on insect behavior is anticipated, let us now return to the topic of death feigning.

5.22 Dopamine, Dopamine, Dopamine

During the 2010s, the term "next-generation sequencing" (NGS) became widely prevalent within the genomics research community. Although initially somewhat hesitant, I was subsequently contacted by Professor Sasaki, who had since relocated to Tamagawa University. The Tokyo University of Agriculture houses the Centre for Genome Analysis of Biological Resources, which announced an open call for collaborative research proposals focused on insect genome sequencing. In early summer 2015, I was invited to submit a research proposal investigating genomic comparisons between long and short strains. Notably, the complete genome of *Tribolium castaneum* had already been sequenced by an American consortium in 2008 (*Tribolium* Genome Sequencing Consortium 2008). This presented a timely opportunity to engage in research targeting the genetic basis of thanatosis (death-feigning behavior).

Next-generation sequencing (NGS) involves the extraction of genomic DNA from an organismal sample, followed by the preparation of a uniformly sized and precisely fragmented DNA library. These DNA fragments are then sequenced using next-generation sequencing platforms. The sequencing instruments process a large number of DNA fragments, ultimately generating sequence data. However, as I only provide the samples, I do not have direct involvement with the instrument's operational procedures. The output from the sequencer consists of DNA fragment sequences, referred to as "reads." The challenges arise at this stage: all subsequent data processing—assembling these reads into contiguous sequences—is performed computationally with the assistance of specialized bioinformatics software and human expertise. This step is technically demanding and requires familiarity with bioinformatics tools, a skill set I do not possess.

In *Tribolium castaneum*, whose complete genome sequence is available, resequencing analysis can be conducted by mapping sequencing reads to the established reference genome for comparative purposes. This represents a significant advantage. The time had finally arrived to leverage the published genome sequence of *T. castaneum* for further research.

Dr. Sasaki and I submitted an application for this project. Since the phrase "analysis of genes for feigning death" seemed inappropriate for a proposal from an agricultural university, we titled the project "Molecular-level investigation of phylogenetic differences in antipredator avoidance behavior in stored-product pest beetles." This title accurately reflected the scope of our research. Indeed, the death-feigning behavior of *Tribolium castaneum* also poses important challenges in the context of pest management.

The submitted project proposal was successfully accepted. On the afternoon of November 2, 2015, I traveled to Kyodo, Tokyo, where the Genome Analysis Centre of Tokyo University of Agriculture is located, to attend a research meeting. This marked the beginning of a collaborative research project with a professor from the Genome Analysis Centre. As planned, I sent samples of the short and long strains of *Tribolium castaneum* to Dr. Sasaki at Tamagawa University. Dr. Sasaki extracted RNA from these samples and forwarded it to the Genome Analysis Centre. Researchers at the centre then performed RNA sequencing (RNA-seq) analysis on the brain tissues using next-generation sequencing technology. In essence, they comprehensively quantified messenger RNA (mRNA) expression levels in the brains and compared the transcriptomes between the short and long strains. This approach, known as transcriptome analysis, leverages sequencing to provide a global overview of gene expression patterns and elucidates the regulatory relationships between RNA functional expression and underlying gene activity.

Transcriptome analysis identified a substantial number of differentially expressed genes—518 in total—between the long and short strains exhibiting death-feigning behavior (Uchiyama et al. 2019). Although both strains were originally derived from the same population, over a decade of selective breeding based on the duration of death feigning resulted in more than 500 genetic expression differences. Crucially, we discovered the key finding: pronounced differential expression was observed, particularly among genes related to dopamine within the tyrosine metabolic pathway. Moreover, numerous genes involved in the broader tyrosine metabolic pathway, which encompasses dopamine metabolism, exhibited significantly divergent expression levels between the strains.

5.23 It Was Dopamine, After All!

Quantitative PCR (qPCR) analysis was subsequently performed to assess the relative expression levels of several tyrosine-metabolizing enzyme genes that exhibited differential expression between the strains, thereby comparing gene expression levels across individuals. The results confirmed that, as anticipated, the expression of dopamine-related genes within the tyrosine metabolic pathway differed significantly between the strains. This study represents the first demonstration worldwide of behavioral differences in death-feigning versus non-death-feigning phenotypes being elucidated at the genomic level, specifically implicating dopamine expression in the brain.

As this was a world-first discovery, we promptly published a series of analyses in 2019 in the open-access journal *Scientific Reports*, a multidisciplinary scientific journal. The paper was titled "World's First Identification of a Gene Cluster Controlling Death Feigning: Elucidating the Mechanism of Thanatosis, a Behavior that Also Captured Fabre's Attention." Since no prior studies had investigated adaptations related to death feigning, it is unsurprising that our work represents pioneering research, given that no one else had undertaken such a specialized investigation into the genetic basis of this behavior.

5.24 The Trajectory of Beetle's Steps

While conducting genetic analyses on the genes controlling death-feigning behavior, I was invited to collaborate with a group of engineering researchers. This opportunity arose through a seminar I organized at Okayama University, which I hosted. In 2007, during an event we called the *Entomology Saturday Seminar*, an ant researcher from Belgium presented his work. At the reception following the seminar, two participants—one an ant researcher and the other an engineer specializing in robotics—connected and subsequently took the initiative to develop a mechanical device capable of measuring insect walking trajectories. This led to the formation of a joint research group focused on free-moving insect locomotion analysis. Around 2015, I was invited to join this collaborative research effort.

Around 2004, shortly after the establishment of the *Tribolium castaneum* strains, I observed that adults of the short-strain beetles appeared to exhibit superior locomotor activity compared to those of the long-strain beetles. However, subjective observation alone was insufficient for scientific publication; it was necessary to quantitatively measure the walking performance. This proved challenging. Initially, I asked students to estimate walking distance by placing the beetles on paper and visually tracking how far they moved. However, since the beetles did not walk in straight lines, accurate distance measurement was not feasible. Consequently, I sought assistance from a researcher at the Ministry of Agriculture, Forestry and Fisheries (MAFF), with whom we had collaborated during our time in Okinawa, and arranged to borrow a device known as the Colour Tracker. This system places insects individually in small Petri dishes, records their walking trajectories via video, and analyzes the paths using specialized PC software to calculate the total distance traveled over a defined period. Utilizing this technology, we were able to quantitatively demonstrate that adults of the short strain exhibited significantly greater walking activity than those of the long strain. These findings were published in *Animal Behaviour* in 2008 (Miyatake et al. 2008) (Fig. 5.4).

However, it was later observed that when the insects reached the edge or corner of the Petri dish, they tended to remain there. In other words, the beetles exhibited a preference for staying in corners, often stopping once they reached these areas. Although the Colour Tracker enabled comparison of walking distances within a fixed time frame, it became apparent that this method could not fully capture the detailed behavioral differences between short- and long-strain adults—differences that are intuitively apparent through direct observation. Specifically, these differences pertain more to walking style rather than simply distance traveled.

Beetles from the non-death-feigning group were consistently active and moved restlessly, whereas those from the death-feigning group typically remained still and moved very little. However, when the latter did initiate movement, their walking patterns appeared distinctly different from the former. Previous studies measured walking distance and activity by confining insects within Petri dishes, but since *T. castaneum* adults preferentially seek corners for safety and stop there, such measurements fail to reflect the beetles' natural directional choices when allowed free movement.

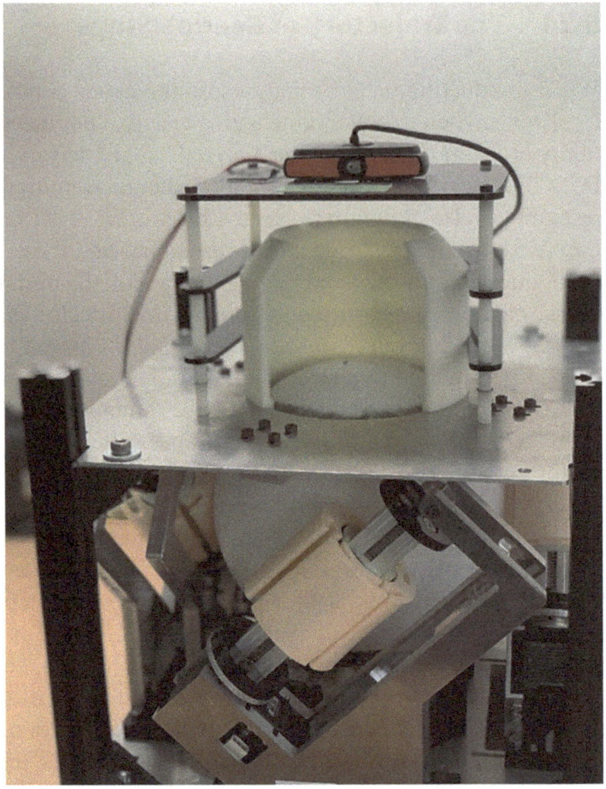

Fig. 5.4 A device for measuring insect walking trajectories. The left panel shows an overview of the measurement setup, in which 16 insects are recorded from above using a video camera. The right panel depicts the walking trajectories of individual insects placed separately in 16 square containers, along with a visualization of the distance walked per unit time, generated using specialized software (Image Hyper II) and displayed on a computer monitor

It was around this time that I received an invitation to collaborate with a group of young engineering researchers who were analyzing the gait trajectories of ants using a custom-built insect treadmill called ANTAM (see Nagaya et al. 2017). I attended the research meeting they organized, and during the subsequent social gathering, our discussion naturally evolved toward the possibility of measuring the walking trajectories of the selected strains of *Tribolium castaneum* adults. Having experienced the COVID-19 pandemic, I can now appreciate that collaborative research often flourishes in informal settings fueled by conviviality—and that such interactions can sometimes lead to unexpected and innovative developments. I am sure I am not alone in feeling this way (Fig. 5.5).

To reiterate, ANTAM is a device developed by Dr. Ryusuke Fujisawa (Kyushu Institute of Technology) and Dr. Naohisa Nagatani (Kyoto Sangyo University) in the field of engineering for recording the walking trajectories of ants and other small insects. It functions as an insect treadmill system (Nagaya et al. 2017). The device consists of a spherical ball on which the insect is placed. A small video camera is mounted above the sphere, and the sphere's movement is controlled by a computer program that responds to the insect's movements. When the insect walks forward, the program rotates the sphere backward; when the insect walks backward, the sphere rotates forward. This mechanism ensures that the insect remains on the top

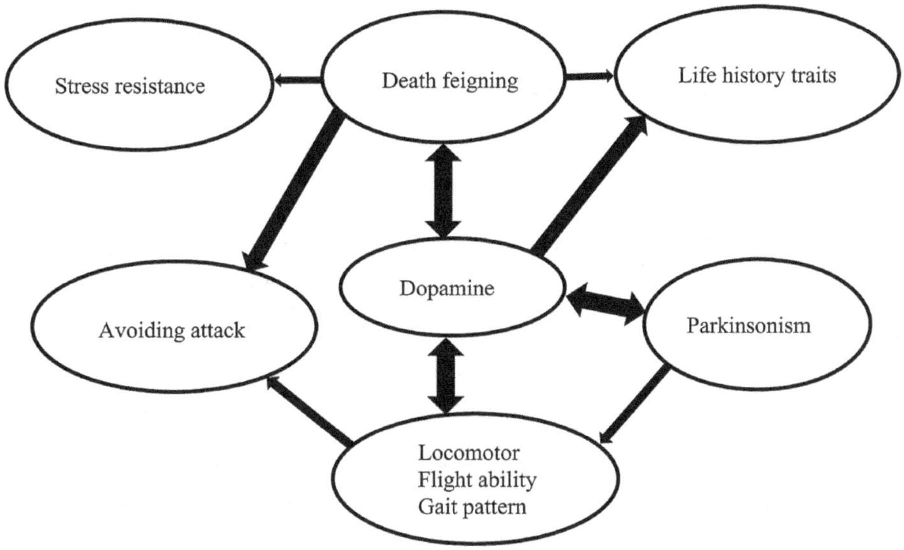

Fig. 5.5 The gait trajectories of ants using a custom-built insect treadmill called ANTAM. (see Nagaya et al. 2017)

of the sphere at all times. The device enables insects to freely walk in place while their walking trajectories are continuously tracked and recorded in real time.

It was not until early 2016 that a new collaboration with engineering researchers commenced, enabling the measurement of gait trajectories of adult beetles from the short and long strains walking on the insect treadmill. Initial data exported to Excel confirmed that short-strain beetles exhibited superior walking performance compared to long-strain beetles, even under conditions of free walking. However, despite visually observing apparent differences in the shapes of the gait trajectories between the strains, I was unable to quantify the "walking style" or pattern based on these measurements.

At that time, I learned that the Ministry of Education, Culture, Sports, Science and Technology (MEXT) had announced an open call for the project titled "Biological Migration Informatics—System Science of Biological Navigation," which aims to integrate engineering and ecology to analyze biological migration as information. I considered that participating in this project might provide opportunities for more precise analysis of gait trajectory data. This was in September 2016.

5.25 Walking at Different Angles

I promptly submitted an application for this major project under the title "Genomic Behavioral Ecology Controlling Navigation Ability," with the aim of advancing research on the migratory and dispersal behaviors of *Tribolium castaneum*. The proposal was successfully accepted, and I was granted the opportunity to participate in the project for 2 years, from 2017 to 2018, for which I am deeply grateful. Through this project, I was able to connect with researchers studying organismal movement and dispersal in natural environments, specialists in biologging—the technique of attaching tracking devices to organisms to monitor their trajectories—and experts in information systems who systematically analyze movement data. This collaboration provided me with many valuable new insights.

The project team included Dr. Takuya Maekawa, a professor of information processing at Osaka University and a specialist in deep learning (artificial intelligence or AI). Dr. Maekawa proposed applying AI techniques to analyze the gait trajectory data of the short and long strains of *Tribolium castaneum*, to which I readily agreed. We subsequently held several meetings to discuss the characteristics of the strains and their trajectories, preparing the data for AI-based analysis. As a result, it became clear that insights gleaned from the trajectory data could not have been obtained through visual inspection alone. The adults of the short and long strains exhibited distinct behaviors in terms of directional changes during locomotion, specifically in their turning or bending patterns. Short-strain adults changed direction intermittently while walking, whereas long-strain adults moved more slowly but tended to maintain straighter paths with minimal angular deviation.

These findings, together with gait trajectory analyses of birds, bears, and rats, were published in the electronic journal *Nature Communications* (UK) in 2020, demonstrating the effectiveness of AI-based learning for analyzing animal movement trajectories (Maekawa et al. 2020).

5.26 Turning Does Not Result in a Reduction in Speed

In this MEXT project, participating researchers convened twice annually to present their findings through poster sessions. During these meetings, Dr. Maekawa observed a common phenomenon across presentations involving humans, mice, nematodes, and *Tribolium castaneum*. Specifically, published data existed on the migratory and dispersal behaviors of dopamine-depleted versus non-deficient individuals across these four species. Leveraging his expertise in artificial intelligence, Dr. Maekawa developed a neural network capable of integrating gait data from multiple species to extract behavioral features distinguishing between normal and dopamine-deficient (pathological) individuals—despite the impossibility of identifying species solely from movement trajectories. Applying this AI technology to analyze gait trajectory data, they discovered that dopamine-deficient mice, nematodes, and *T. castaneum* beetles all exhibit shared motor deficits, including an impaired ability to decelerate smoothly prior to turning.

How should we interpret the gait disturbance observed in the long strain of *Tribolium castaneum*, which exhibits prolonged death-feigning behavior? Dr. Maekawa visited Okayama University several times to discuss the data with me and the graduate student who collected the gait trajectories. Together, we repeatedly compared the gait trajectory data with video recordings of individuals walking on the treadmill. From these analyses, we hypothesized that the unusual walking pattern of the long strain results from an impaired ability to decelerate smoothly before turning. This hypothesis was successfully validated using artificial intelligence techniques.

Our findings suggest that the mechanisms by which dopamine deficiency affects locomotion are evolutionarily conserved across species, from nematodes and insects to humans. These results were published in the British scientific journal *Nature Communications* (online), accompanied by a press release in 2021 (Maekawa et al. 2021).

5.27 Genomes Controlling Death-Feigning Duration

On the other hand, significant progress was made in genome analysis in 2021. The genome, which comprises the complete genetic information of living organisms, including humans, is often referred to as the "blueprint of life." The mRNA expression profiles of short and long strains of *Tribolium castaneum* were analyzed and reported in 2019 (Uchiyama et al. 2019). Subsequently, collaborative research between the Genome Research Center at Tokyo University of Agriculture (NODAI) and Tamagawa University advanced further. When genomic information for a given species is publicly available, it can serve as a reference for genome decoding. This approach, known as DNA resequencing, was mentioned earlier. Since the entire genome sequence of *T. castaneum* was published in 2008, this technique allows us to fully leverage the available genomic resources for more detailed analyses.

Our team—although the DNA sequencing and data compilation were carried out almost exclusively by a researcher from Tokyo University of Agriculture—successfully obtained whole-genome sequences for both the short and long strains of *Tribolium castaneum*. We compared these sequences to those of a non-artificially selected beetle available in public databases. The analysis revealed that the long-strain beetles, which exhibit prolonged death-feigning behavior, harbored numerous mutations in dopamine-related genes. Additionally, mutations were identified in genes involved in caffeine metabolism, cytochrome-related oxidoreductase functions (heme proteins involved in redox reactions), lifespan regulation, and circadian rhythm control systems—commonly referred to as the "body clock." Compared to the reference genome, the long strain showed a higher number of mutations affecting sleep–wake cycles and hormonal regulation than the short strain, which tends to escape rather than feign death. These findings suggest that genetic differences influencing neurochemical pathways and physiological rhythms may underlie behavioral variation in death feigning (Fig. 5.6).

This outcome was also anticipated. Since we selected for the duration of death-feigning behavior in our breeding program, it is unsurprising that these behavioral and temporal traits are influenced by multiple physiological mechanisms within the organism. Specifically, while the tyrosine metabolic pathway—including dopamine degradation—plays a key role, mutations were also observed across various other genomic regions and gene pathways, particularly in the long strain compared to the short strain. This result aligns with our initial expectations, as the prolonged

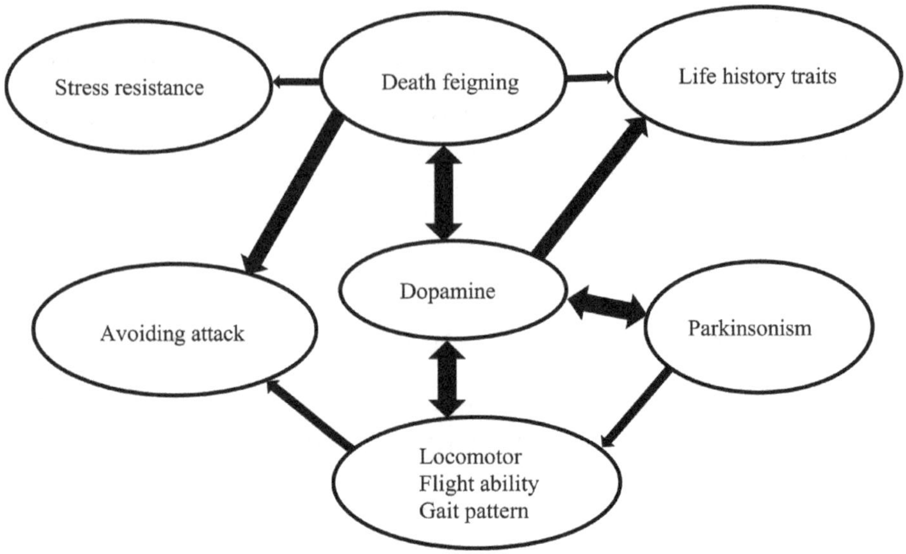

Fig. 5.6 Schematic illustration of death-feigning syndrome

death-feigning phenotype is unlikely to be governed by a simple Mendelian inheritance pattern involving a single gene. Instead, we hypothesized from the outset that this trait is polygenic, controlled by multiple genes. This form of inheritance, where multiple genes collectively influence a trait, is referred to as quantitative trait inheritance.

These genes are all critical in determining the life history traits of insects—that is, the timing of various physiological and developmental processes throughout their lifespan. Although our selective breeding targeted only the duration of death-feigning behavior, numerous genes associated with different aspects of an organism's life timing were altered. We published these findings in the open-access journal *Scientific Reports*, published by the UK-based Nature Publishing Group (Tanaka et al. 2021). An important insight emerging from this work, which points to future research directions, is that the prevalence of mutations in these genes among long-strain beetles that feign death for extended periods may have implications for understanding human diseases.

The findings from the genetic analyses were reflected in the observable phenotypes. In addition to elucidating molecular mechanisms, researchers in the laboratory investigated the mode of inheritance of phenotypic traits. Specifically, the segregation patterns of traits such as death-feigning duration and activity levels were examined in F1 offspring (resulting from crosses between long and short strains) and in F2 males and females (the subsequent generation).

The observed segregation patterns deviated significantly from classical Mendelian inheritance, confirming that multiple genetic factors contribute to the duration of death feigning as a visibly inheritable trait (Matsumura and Miyatake 2022). Correspondingly, DNA resequencing data corroborated these findings at the genomic level, supporting the polygenic nature of this quantitative trait. However, phenotypic analyses also highlighted an unresolved question: which specific groups of genes, categorized by their biological functions, are involved in the regulation of death-feigning behavior?

The behavior of movement is a fundamental trait critical to the survival of living organisms. Many survival-related traits, including defense mechanisms and reproductive strategies, depend on both the act of movement and the decision to remain still, each governed by specific modes of inheritance. Moreover, at both the molecular level and in observable behavioral phenotypes, death feigning represents a complex trait inherently linked to the dichotomy of moving versus not moving.

5.28 A Proposal for Death-Feigning Syndrome

Let us summarize the findings from our experiments investigating the mechanisms underlying death-feigning behavior in insects. The study of death feigning has revealed a complex phenomenon that we propose to term "*death-feigning syndrome*." This syndrome is rooted in genetic factors that regulate organismal movement and can be conceptualized as follows:

Feigning death ⇔ activity level ⇔ dopamine expression in the brain ⇔ arousal mediated by dopamine ⇔ tyrosine metabolic pathway ⇔ homologues of Parkinson's disease-related genes.

Some aspects of this phenomenon have previously been described as a behavioral syndrome associated with death feigning (Nakayama and Miyatake 2010). Therefore, we adopt the term *death-feigning syndrome* to encompass this broader framework.

Similar patterns linking movement and death-feigning behaviors have been observed in other insects, such as the bean beetle *Callosobruchus chinensis*. In this species, individuals that frequently feign death exhibit poor flight ability, whereas those that do not feign death are proficient flyers. This observation may represent another manifestation of the death-feigning syndrome.

5.29 When Should I Wake Up from Death Feigning?

When I was working as an official in Okinawa Prefecture, a senior colleague knowledgeable about insects and studying the ecology of sweetpotato weevils told me that these weevils were being transported by ants while feigning death. Of course, remaining in a state of death feigning indefinitely would be risky. At some point, the insect must terminate this behavior and resume activity. The question of *when* to awaken from death feigning is thus both a longstanding and yet still unresolved issue.

It was Dr. Jean-Henri Fabre who first discussed the concept of awakening from death feigning. In his experiments, he stimulated a small ground beetle, *Scarites aterrimus* (the leopard louse beetle), to feign death on a table, then introduced a larger ground beetle, *Scarites sulcatus*, into the environment. Fabre observed that the smaller beetle continued to play dead only while the larger beetle was nearby, but resumed normal activity as soon as the larger beetle withdrew. He proposed that death feigning might simply result from temporary neurological paralysis.

The adaptive significance of death feigning is described in Chap. 2; here, let us focus on the concept of "awakening"—that is, when the prey should terminate death-feigning behavior. From previous experience, I knew that placing a drop of magic ink near the sweetpotato weevil *Cylas formicarius* while it was feigning death would cause it to awaken. I also observed that repeated stimulation, such as repeated poking, reduced the likelihood of death feigning, whereas in some cases, it caused the insect to feign death for longer durations, as if entering an unusually deep state of immobility. However, I was unable to devise a method to quantify the intensity of stimuli required to trigger awakening from death feigning or to express this threshold as a precise numerical value.

At that time, a graduate student conducting experiments with *Tribolium castaneum* on the ANTAM sphere—insect treadmill described earlier—encountered a problem: slight vibrations generated by the ANTAM device frequently caused adult beetles to stumble or fall over. This phenomenon was observed more often in adults of the long strain, which exhibit prolonged death-feigning behavior, than in the short strain.

Initially suspecting a malfunction of the ANTAM equipment, I discussed the issue with Dr. Nagaya, an engineering researcher. During our conversation on social media, Dr. Nagaya suggested that the vibrations themselves might contribute to the long-strain beetles' increased rubbing behavior on the sphere. Intrigued by this hypothesis, a student, a researcher, and I promptly transported the beetles to Kyoto Sangyo University, where Dr. Nagaya is based, to investigate further.

On July 20, 2017, we brought the beetles to Dr. Naohisa Nagaya's laboratory in the Faculty of Computer Science and Engineering at Kyoto Sangyo University. After a train and bus ride, we walked up a slight hill to reach the lab, sweating from the effort. Once there, we unpacked our equipment and notes for the day's experiments. We first placed the beetles in Petri dishes and observed their walking behavior. Next, we prepared Petri dishes equipped with transducers underneath to apply vibrations. We then placed adults of the long strain into these vibrating Petri dishes and began our observations.

After some time, an interesting phenomenon occurred. Adult beetles of the long strain were placed in a Petri dish equipped with an oscillator underneath, delivering a weak vibratory stimulus at approximately 20 Hz. Under this low-frequency vibration, the beetles, while still feigning death, slid sideways across the surface of the Petri dish. However, when the vibration frequency was increased to 50 Hz, the long-strain adults suddenly terminated their death-feigning behavior and began to walk actively. To confirm this observation, I tested four adult beetles that I had brought with me. Each individual consistently awoke and started walking upon exposure to the 50 Hz vibration stimulus.

This is!

Using a long strain characterized by a pronounced depth of death feigning allows us to quantify the awakening from this behavior. To be more specific, when insects are collected from the field and induced to feign death, applying a stimulus of a given intensity does not guarantee that the stimulus itself is the critical factor triggering arousal. It is also possible that individual variation in the depth of death feigning causes some insects to awaken simply because their immobility was shallow, coincidentally responding at the moment of stimulation. However, by using long strains with consistently deep death-feigning responses, this variability can be minimized, enabling a more precise and straightforward verification of the relationship between stimulus intensity and awakening.

So, how do you adjust the intensity of the vibration? Dr. Nagaya, an engineer, replied, "That's easy," and asked a graduate student at Kyoto Sangyo University, who was in the same room, "Can I borrow your amplifier for a moment?" By varying the amplitude of the sound waves transmitted from the amplifier, it is possible to easily modulate the vibration intensity of the Petri dish carrying the insects. That's engineering—brilliant!

The previously mentioned 20 Hz and 50 Hz vibrations were generated using this method. By measuring the displacement amplitude of the Petri dish vibrating vertically, we could precisely quantify the strength of the vibrations experienced by the insects.

As the saying goes, "A rice cake is a rice cake shop" (i.e., experts excel in their field), but having studied animal behavior, this was a development I would never have conceived. It remains crucial to engage directly with experimental setups or fieldwork when uncertainties or problems arise—a principle that held true even before the COVID-19 pandemic.

We have established a long strain of *Tribolium castaneum* that consistently feigns death for an hour or more, even in response to minimal stimuli, as well as a short strain that does not exhibit death-feigning behavior regardless of the intensity of stimulation. Moreover, the long and short strains were independently bred from separate populations, maintaining their distinct characteristics throughout the breeding process.

First, the effect of vibration intensity on arousal from death feigning will be quantitatively investigated in *Tribolium castaneum*. Next, by crossing the short and long strains and measuring the vibration amplitude required to induce arousal in the F1 offspring, we aim to determine the dominance or recessiveness of the arousal trait. Furthermore, by analyzing the F2 generation—produced by crossing F1 males and females—we can examine the distribution of vibration intensities required for arousal and thereby characterize the genetic architecture underlying this trait. At that time, we also maintained three independent long and short breeding strains each in the closely related species *T. confusum* and *T. freemani*. This led to the conception of an additional research question: Do closely related species differ in the vibration intensity thresholds required for arousal from death feigning?

Thus began a new chapter of interdisciplinary research combining engineering and animal behavior. On the day I traveled to Kyoto to investigate why insects were stumbling on the ANTAM device, an unexpected conversation with an engineering researcher sparked the development of a project that eventually led to the publication of two international papers—turning a small seed of an idea into substantial outcomes. On the way back from Kyoto Sangyo University, I couldn't help but celebrate with the students and colleagues who accompanied me at a traditional Japanese pub in Kitaoji, where we got off the bus.

5.30 Arousal from Death Feigning: Vibration Stimulus

Upon returning to Okayama University, we promptly assembled the vibratory apparatus and prepared adult beetles from our laboratory-maintained strains for the experiment. While the sound generated by the amplifier had posed no significant issue on the engineering campus, it produced a considerably louder vibratory noise in our agricultural science facilities. To minimize acoustic disturbance to adjacent laboratories, we introduced rubber sheets to absorb the vibrations. The experiment was subsequently initiated using adult beetles of the long-duration strain. Together with the students, I carefully positioned the transducers on the rubber sheets, tempering our excitement as we embarked on a novel experimental procedure.

An adult *Tribolium castaneum* beetle from the long-duration strain was gently placed in a Petri dish positioned atop the oscillator in order to apply vibratory stimulation intended to elicit thanatosis (death-feigning behavior). Each experimental trial was conducted by a team of three individuals, including myself. One person was responsible for handling the beetles, another for controlling the vibration parameters, and the third for recording observational data. Vibratory stimuli were applied incrementally, with frequencies increased stepwise from 25 Hz to 40 Hz, 50 Hz, and finally 60 Hz.

For each applied frequency, the vertical amplitude of the Petri dish was measured, and the amplitude was used as a proxy for vibration intensity. The relationship between vibration amplitude and the rate of arousal from thanatosis was investigated. Results indicated that adult beetles of the long-duration strain resumed movement when the amplitude exceeded 0.14 mm (corresponding to approximately 40 Hz), and that nearly all individuals awoke when the amplitude exceeded 0.2 mm. Interestingly, when the vibration frequency was further increased to approximately 100 Hz, no beetles exhibited arousal from thanatosis.

These results demonstrated that a specific threshold range of vibratory stimulation is required to induce arousal from thanatosis, suggesting the existence of an optimal vibration range for recovery from this immobile state. To investigate whether this arousal threshold has a genetic basis—i.e., whether susceptibility to vibratory-induced arousal is heritable—we conducted a series of crossing experiments. Males and females from both the short- and long-duration strains were interbred to produce F1 offspring, which were subsequently crossed to generate F2 (grandchild) populations. The arousal thresholds in response to vibratory stimulation were then systematically examined in these F2 individuals. In total, more than 500 adult beetles were tested in stationary observational experiments.

In the F2 population, the stimulus intensity required to elicit arousal from thanatosis varied continuously, ranging from as low as 0.001 mm to over 0.5 mm in amplitude. Furthermore, a clear positive correlation was observed between the amplitude threshold required for arousal and the duration of thanatosis. These findings were compiled and published in the international journal *Behavior Genetics*, in a paper entitled "Arousal from tonic immobility by vibrational stimulus" (Miyatake et al. 2019).

Are the vibration thresholds required for arousal from thanatosis conserved across insect species? To address this question, we conducted parallel investigations using *Tribolium confusum* (Nakayama and Miyatake 2010) and *Tribolium freemani* (Konishi et al. 2020), two closely related species in which we had successfully established both short- and long-duration strains through artificial selection, following the same methodology used for *T. castaneum*. In these species, we specifically examined the vibratory stimulus intensity required to induce arousal in adult beetles of the long-duration strains, which had been genetically selected for prolonged thanatosis.

Are the vibratory thresholds required to induce arousal from thanatosis conserved across different insect species? To investigate this, we successfully established short- and long-duration strains of *Tribolium confusum* (Nakayama and Miyatake 2010) and *Tribolium freemani* (Konishi et al. 2020)—a closely related species—using the same artificial selection methods applied to *T. castaneum*. We subsequently examined the vibration intensities necessary to elicit arousal from thanatosis in adult beetles of the long-duration strains, which had been selectively bred to exhibit prolonged death-feigning behavior.

Tribolium confusum was aroused from thanatosis by longitudinal vibrations with amplitudes of approximately 0.25 mm, whereas *T. freemani* required slightly higher amplitudes of around 0.3 mm. The underlying mechanisms and the adaptive significance of these interspecific differences in vibratory arousal thresholds remain unclear. Notably, the death-feigning duration of nonselected *T. freemani* is approximately 3000 seconds, which is substantially longer than the approximately 300 seconds observed in *T. confusum* and the 10–30 seconds typical of *T. castaneum*. Given that these differences may reflect species-specific variation in the depth or intensity of tonic immobility, we published the findings of this comparative interspecies study in the *Journal of Ethology*, the official publication of the Japanese Society for Ethology, in 2020 (Ishihara et al. 2021).

The duration of thanatosis varies considerably among insect species. For example, individuals of the sweetpotato weevil *Cylas formicarius*—discussed in the previous chapter of this book—when collected from field populations, typically exhibit death-feigning behavior lasting from a few seconds to several minutes. In contrast, another invasive species attacking sweet potatoes, the West Indian sweetpotato weevil *Euscepes postfasciatus*, displays markedly prolonged thanatosis, maintaining immobility for tens of minutes to over an hour (T.M., personal observation). Notably, the duration of thanatosis in field populations of *Tribolium castaneum* generally ranges from approximately 1–6 minutes, although this varies depending on the geographic origin of the population.

Fabre also measured the duration of thanatosis in several beetle species, noting in his seminal work *Souvenirs Entomologiques* that this duration varies considerably among species (Fabre 1910). What accounts for the interspecific variation in thanatosis duration, and how is this related to the intensity of vibratory stimuli required to induce arousal from the death-feigning state? The adaptive significance of these differences likely cannot be fully understood without comprehensive investigations into the principal predator species in natural habitats, as well as detailed studies on the duration of thanatosis and arousal thresholds of both predators and prey. How are the duration of thanatosis and the timing of arousal regulated in field populations? This remains an open question for future research. The following section presents examples of previous studies that have documented and discussed variations in thanatosis duration under natural conditions.

5.31 Arousal from Death Feigning: Pheromone

Pheromones serve as critical signaling molecules that influence survival and reproductive success in many organisms. The effectiveness of death feigning as an anti-predation strategy may be modulated by additional factors related to survival and reproduction beyond mere predator avoidance. Two such factors are foraging and mating. Even if predation is successfully avoided, an individual that fails to acquire sufficient food cannot survive. Similarly, in sexually reproducing organisms, avoidance of predation alone does not ensure fitness if successful fertilization does not occur. In both scenarios, the adaptive value of prolonged thanatosis may be nullified, favoring individuals that terminate death-feigning behavior and switch to alternative tactics to enhance their fitness.

Considering aggregation pheromones specifically, in *Tribolium castaneum*, males release pheromones that attract both male and female adults. The adaptive significance of this pheromone release is generally attributed to enhanced predator avoidance and efficient resource localization. Alternatively, it is possible that male-produced pheromones originally evolved to attract females, thereby allowing males capable of detecting these signals to locate mates more rapidly. In either case, individuals that terminate death feigning in response to such chemical cues and adopt alternative behaviors may gain a selective advantage.

We first exposed adults of the long-duration strain, selectively bred to exhibit prolonged thanatosis, to artificially synthesized aggregation pheromones and found that both female- and male-derived aggregation pheromones effectively induced arousal from death-feigning behavior (Ishikawa et al. 2024). Since *Tribolium castaneum* does not produce sex pheromones, the effects of sex pheromones on arousal from thanatosis were investigated using *Cylas formicarius*. Himuro and Miyatake (2024) conducted two experiments with *C. formicarius*. When death-feigning males and females were exposed to sexually mature and unmated females, sexually immature and unmated females, sexually mature males, or control conditions (no exposure), only death-feigning males exposed to sexually mature and unmated females exhibited significantly accelerated arousal. Exposure to other stimuli did not affect arousal latency compared to controls, and females did not exhibit arousal in response to any treatment. Further, exposure to artificially synthesized sex pheromones elicited immediate arousal in death-feigning males but not females. These results indicate that sex pheromones have a rapid arousing effect on thanatosis, but this effect is sex-specific, occurring only in the sex that receives the signal.

5.32 Are There Regional Differences in Death?

In Chap. 3, it was noted that the sweetpotato weevil, *Cylas formicarius*, exhibits thanatosis in a conspicuous posture with its legs extended. Beetles generally display two primary types of death-feigning postures. The first type involves the individual turning onto its back upon stimulation and retracting its legs tightly against the abdomen, resulting in immobilization. This posture is commonly observed in click beetles (family Elateridae), which are well-known for their death-feigning behavior. Similarly, stag beetles, bean beetles (*Tribolium* spp.), and many weevil species adopt this pose. It is hypothesized that this posture has evolved as a form of camouflage: by folding the legs close to the body, the beetle reduces its shadow and blends more effectively into the substrate.

The second type of death-feigning posture is characterized by legs extended away from the body and is observed in species such as *Cylas formicarius* and various longhorn beetles. The functional significance of this leg extension during thanatosis remains unclear. Some hypotheses propose that extending the legs may reduce the likelihood of being swallowed by predators. Alternatively, it has been suggested that insects perched on branches may fall more quickly when immobile if their legs are extended, possibly facilitating rapid escape from predation.

It has long been recognized that these two types of death-feigning postures vary among species. However, in 2021, researchers at Nagasaki University reported geographic variation in death-feigning posture in the ferocious water bug, *Appasus japonicus* (Ohba and Matsuda 2021). Specimens collected from Hokkaido, Hyogo, Nagasaki, and Kagoshima exhibited a mixture of leg-retracting and leg-extending types, whereas all individuals collected from Ibaraki displayed the leg-extending posture exclusively. While differences in predator communities and environmental conditions are thought to influence death-feigning postures, studies addressing geographic variation in thanatosis remain in their infancy.

As noted in Chap. 3, the spiny grasshopper exhibits a unique defensive posture when ingested by frogs: its spiny hind legs are erected vertically. The thoracic spines, present since adulthood, extend horizontally, while the stiffened, vertically erect hind legs resemble the shape of a water caltrop, as implied by its name. This posture renders the grasshopper difficult for frogs to swallow, effectively serving as a physical defense mechanism (Honma et al. 2006). Notably, this rigid posture is not assumed in response to attacks by birds; it is specifically triggered upon ingestion by frogs. Thus, this strategy functions as a swallow-avoidance mechanism against a particular predator.

From this perspective, the death-feigning posture of *Cylas formicarius*, which involves pinning the hind legs backward, may hold functional significance. However, because the legs of this ant-associated weevil are relatively soft, it remains unclear whether they serve a similar defensive function.

5.33 Is Random Death-Feigning Duration Optimal?

In 2020, Dr. Franks and colleagues from the University of Bristol, UK, conducted a study on thanatosis in antlion larvae collected from the Island of Guernsey, a British Crown dependency located in the English Channel between France and southern England. Antlions are voracious predators whose larvae construct pitfall traps in sandy substrates to ambush and capture vulnerable prey. Upon seizing prey, the larvae immobilize and pierce them with their sharp mandibles, digesting the prey internally. Interestingly, these predatory larvae also exhibit death-feigning behavior when threatened by potential enemies, sometimes remaining immobile for over an hour.

The nests of antlion larvae are often situated in open areas exposed to soaring birds, making them susceptible to avian predation. Paradoxically, their primary anti-predator strategy is a highly passive one—thanatosis.

While weighing antlion larvae during ecological fieldwork, Franks and colleagues observed that the larvae would abruptly freeze and cease movement. Using a stopwatch, they measured the duration of immobility, finding substantial variation ranging from a few seconds to as long as 61 minutes. They interpret this behavior not simply as an act of deceiving predators but rather as a form of concealment by "hiding in plain sight." Accordingly, they coined the term "post-contact immobility" to describe this insect death-feigning behavior (see Chap. 2).

An intriguing aspect of their findings was that when individual antlion larvae were stimulated to feign death twice, the durations of the two thanatosis episodes varied considerably. This variability led the researchers to hypothesize that the seemingly unpredictable durations of death feigning may represent an adaptive survival strategy, enhancing the larvae's ability to evade predators.

They proposed that if the duration of thanatosis were predictable, predators might learn and exploit this pattern. For example, birds may occasionally drop larvae that have been excavated from the sand. To investigate this hypothesis, simulations were conducted in collaboration with computational modelers to assess whether unpredictable durations of death feigning would confer a selective advantage by reducing the likelihood of repeated attacks on the same individual.

Simulation results demonstrated that antlion survival was significantly compromised when larvae exhibited short durations of thanatosis, whereas prolonged durations were less effective in enhancing survival. Notably, the highest survival rates were observed when the duration of death feigning was unpredictable. Based on these findings, it was concluded that natural populations exhibiting variable and unpredictable thanatosis durations are more likely to achieve higher survival, particularly in prey species that aggregate in groups. These results were published in the *Proceedings of the Royal Society B* in 2020 (Sendova-Franks et al. 2020) and *Biology Letters* in 2021 (Franks et al. 2024).

References

Acheampong A, Mitchell BK (1997) Quiescence in the Colorado potato beetle, *Leptinotarsa decemlineata*. Entomol Exp Appl 82:83–89

Asakura M, Matsumura K, Ishihara R, Miyatake T (2022) Freezing or death feigning? Beetles selected for long death feigning showed different tactics against different predators. Ecol Evol 12:e8533

Fabre JH (1910) Souvenirs Entomologiques. Delgrave Paris, Paris

Falconer DS (1989) Introduction to quantitative genetics, 3rd edn. Longman, New York

Franks NR, Worley A, Fortune GT, Goldstein RE, Sendva-Franks AB (2024) Seeking safety: movement dynamics after post-contact immobility. PLoS One 19(8):e0307370

Harano T, Okada K, Nakayama S, Miyatake T, Hosken DJ (2010) Intralocus sexual conflict unresolved by sex-limited trait expression. Curr Biol 20:2036–2039

Hayashi T, Matsumura K, Miyatake T (2024) Body-size-dependent predation by some jumping spider species (Araneae: Salticidae) on *Tribolium castaneum* (Coletptera: Tenebrionidae). Appl Entomol Zool 59:267–271

Himuro C, Miyatake T (2024) Mate search or predation avoidance? Sex pheromone interrupts death feigning of males in the sweet potato weevil *Cylas formicarius*. J Ethol 42:169–176

Honma A, Oku S, Nishida T (2006) Adaptive significance of death feigning posture as a specialized inducible defence against gape-limited predators. Proc Royal Soc B 273:1631–1636

Ishihara R, Mastumura K, Jones JE, Yuhao J, Fujisawa R, Nagaya N, Miyatake T (2021) Arousal from death feigning by vibrational stimuli: comparison of *Tribolium* species. J Ethol 39:107–113

Ishikawa M, Matsumura K, Miyatake T (2024) Aggregation pheromone interrupts death feigning in the red flour beetle *Tribolium castaneum*. J Ethol 42:71–75

Katsuki M, Miyatake T (2009) Effects of temperature on mating duration, sperm transfer and remating frequency in *Callosobruchus chinensis*. J Insect Physiol 55:113–116

Kiyotake H, Matsumoto H, Nakayama S, Sakai M, Miyatake T, Ryuda M, Hayakawa Y (2014) Gain of long tonic immobility behavioral trait causes the red flour beetle to reduce anti-stress capacity. J Insect Physiol 60:92–97

Konishi K, Matsumura K, Sakuno W, Miyatake T (2020) Death feigning as an adaptive anti-predator behaviour: further evidence for its evolution from artificial selection and natural populations. J Evol Biol 33:1120–1128

Maekawa T, Higashide D, Hara T, Matsumura K, Ide K, Miyatake T, Kimura KD, Takahashi S (2021) Cross-species behavior analysis with attention-based domain-adversarial deep neural networks. Nat Commun 12:5519. https://doi.org/10.1038/s41467-021-25636-x

Maekawa T, Ohara K, Zhang Y, Fukutomi M, Matsumoto S, Matsumura K, Shidara H, Yamazaki S, Fujisawa R, Ide K, Nagaya N, Yamazaki K, Koike S, Miyatake T, Kimura K, Ogawa H, Takahashi S, Yoda K (2020) Deep learning-assisted comparative analysis of animal trajectories with DeepHL. Nat Commun 11:5316. https://www.nature.com/articles/s41467-020-19105-0

Matsumura K, Miyatake T (2018) Responses to relaxed and reverse selection in strains artificially selected for duration of death-feigning behavior in the red flour beetle, *Tribolium castaneum*. J Ethol 36:161–168

Matsumura K, Miyatake T (2022) Polygene control and trait dominance in death-feigning syndrome in the red flour beetle *Tribolium castaneum*. Behav Genet 52:298–305

Matsumura K, Yumise K, Fujii Y, Hayashi T, Miyatake T (2020) Anti-predator behaviour depends on male weapon size. Biol Lett 16:20200601. https://doi.org/10.1098/rsbl.2020.0601

Miura Y, Takii H, Matsumura K, Miyatake T (2025) Experimental evolution of prey by predation pressure: death-feigning duration becomes longer in the presence of predators. Appl Entomol Zool 60:257

Miyatake T, Matsumura F (2004) Intra-specific variation in female remating in *Callosobruchus chinensis* and *C. maculatus*. J Insect Physiol 50:403–408

Miyatake T, Katayama K, Takeda Y, Nakashima A, Sugita A, Mizumoto M (2004) Is death-feigning adaptive? Heritable variation in fitness difference of death-feigning behaviour. Proc R Soc Lond Ser B 271:2293–2296

Miyatake T, Tabuchi K, Sasaki K, Okada K, Katayama K, Moriya S (2008) Pleiotropic anti-predator strategies, fleeing and feigning death, correlated with dopamine levels in *Tribolium castaneum*. Anim Behav 75:113–121

Miyatake T, Nakayama S, Nishi Y, Nakajima S (2009) Tonically immobilized selfish prey can survive by sacrificing others. Proc R Soc B 276:2762–2767

Miyatake T, Matsumura K, Kitayama R, Otsuki K, Yuhao J, Fujisawa R, Nagaya N (2019) Arousal from tonic immobility by vibration stimulus. Behav Genet 49:478–483

Moriya S, Miyatake T (2001) Eradication programs of two sweetpotato pests, *Cylas formicarius* and *Euscepes postfasciatus*, in Japan with special reference to their dispersal ability. JARQ 35:227–234

Nagaya N, Mizumoto N, Abe MS, Dobata S, Sato R, Fujisawa R (2017) Anomalous diffusion on the servosphere: a potential tool for detecting inherent organismal movement patterns. PLoS One 12(6):e0177480. https://doi.org/10.1371/journal.pone.0177480

Nakayama S, Miyatake T (2009a) Positive genetic correlations between life-history traits and death-feigning behavior in adzuki bean beetle (*Callosobruchus chinensis*). Evol Ecol 23:711–722

Nakayama S, Miyatake T (2009b) Genetic trade-off between abilities to avoid attack and to mate: a cost of tonic immobility. Biol Lett 6:18–20

Nakayama S, Miyatake T (2010) A behavioral syndrome in the adzuki bean beetle: genetic correlation among death feigning, activity, and mating behavior. Ethology 116:108–112

Nishi Y, Sasaki K, Miyatake T (2010) Biogenic amines, caffeine and tonic immobility in *Tribolium castaneum*. J Insect Physiol 56:622–628

Ohba S, Matsuda I (2021) Discovery of geographical variation of death-feigning in *Appasus japonicus* (Heteroptera: Belostomatidae). Rostria: Trans Hemipterol Soc Japan 66:22–25, 2021-09-30 (in Japanese)

Ohno T, Miyatake T (2007) Drop or fly? Negative genetic correlation between death-feigning intensity and flying ability as alternative anti-predator strategies. Proc R Soc B 274:555–560

Okada K, Miyatake T (2006) Intra-sexual dimorphism in male mandibles and male aggressive behavior in the broad-horned flour beetle *Gnatocerus cornutus* (Coleoptera: Tenebrionidae). J Insect Behav 19:457–467

Okada K, Miyatake T (2010) Effect of losing on male fights of broad-horned flour beetle, *Gnatocerus cornutus*. Behav Ecol Sociobiol 64:361–369

Ouyang H, Wu P, Zhang R, Haseeb M (2020) Trade-off investment between tonic immobility and mate search in the Sweetpotato weevil, *Cylas formicarius* (Coleoptera: Brentidae). Insects 11:774. https://doi.org/10.3390/insects11110774

Ouyang H, Zhang R, Haseeb M (2023) Investment trade-off between mating behavior and tonic immobility in the Sweetpotato weevil *Cylas formicarius* (Coleoptera: Brentidae). Insects 14:73. https://doi.org/10.3390/insects14010073

Ruxton G (2006) Grasshoppers don't play possum. Nature 440:880

Sakai M (2021) Death-feigning in insects, Entomology Monographs. Springer Nature Singapore Pte Ltd

Sendova-Franks AB, Worley A, Franks NR (2020) Post-contact immobility and half-lives that save lives. Proc R Soc Lond B Biol Sci 287:20200881. https://doi.org/10.1098/rspb.2020.0881

Tanaka K, Sasaki S, Matsumura K, Yajima S, Miyatake T (2021) Genomic characterization between strains selected for death-feigning duration for avoiding attack of a beetle. Sci Rep 11:1–18. https://doi.org/10.1038/s41598-021-00987-z

Tribolium Genome Sequencing Consortium (2008) The genome of the model beetle and pest *Tribolium castaneum*. Nature 452:949–955

Uchiyama H, Sasaki K, Hinosawa S, Tanaka K, Matsumura K, Yajima S, Miyatake T (2019) Transcriptomic comparison between beetle strains selected for short and long durations of death feigning. Sci Rep 9:14001. https://doi.org/10.1038/s41598-019-50440-5

Yamane T, Miyatake T (2005) Intra-specific variation in strategic ejaculation according to level of polyandry in *Callosobruchus chinensis*. J Insect Physiol 51:1240–1243

Yamane T, Okada K, Nakayama S, Miyatake T (2010) Dispersal and ejaculatory strategies associated with exaggeration of weapon in an armed beetle. Proc R Soc B 277: 1705–1710

Yanagi S, Miyatake T (2002) Effects of maternal age on reproductive traits and fitness components of the offspring in the bruchid beetle, Callosobruchus chinensis (Coleoptera: Bruchidae). Physiol Entomol 27:261–266

Yanagi S, Miyatake T (2003) Costs of mating and egg production in female *Callosobruchus chinensis*. J Insect Physiol 49:823–827

Yuhao J, Ryuji Y, Matsumura K, Miyatake T (2021) Effects of caffeine on mating behavior and sperm precedence in *Tribolium castaneum*. Ethology 127:45–49

Death-Feigning Behavior in Humans and Its Links to Disease

6

Abstract

The connection between death feigning and human behavior is a subject for future exploration. In this chapter, I will briefly introduce what is currently known about immobility in animals—that is, death feigning—and its relationship to human activity. In particular, I will provide an overview of how research on death feigning relates to Parkinson's disease, posttraumatic stress disorder (PTSD), sexual assault, thanatology, and its potential applications to human industrial activities. Why do humans feign death? This question points to the possibility that, throughout the vast course of human evolutionary history, the behaviors of death feigning and immobility have persisted—not eliminated by natural selection—because they served as survival strategies.

Keywords

Posttraumatic stress disorder · Sexual assault · Thanatology · Parkinson's disease · Mental illness · Psychiatric disorder · Hysteria

Numerous historical records document instances of death feigning in humans. For example, during the Sengoku period in Japan, many local samurai reportedly survived by exhibiting thanatosis, or death-feigning behavior. Similarly, in the Chinese historical novel Romance of the Three Kingdoms, there is an account of a general who survived by feigning death (Yoshikawa 1989). From antiquity to the present, and across diverse cultures worldwide, death-feigning behavior is extensively reported. Importantly, this behavior is not exclusive to humans; many animal species also display thanatosis. Over evolutionary time, numerous taxa have independently developed and refined this adaptive antipredator strategy to enhance survival probabilities.

© The Author(s), under exclusive license to Springer Nature Singapore Pte Ltd. 2026

T. Miyatake, *Death Feigning*, Ecological Research Monographs,
https://doi.org/10.1007/978-981-95-5319-8_6

However, historical accounts cannot be strictly regarded as scientific evidence. In the context of human death-feigning behavior, terms such as thanatosis, tonic immobility, and catalepsy (rigidity) are commonly employed. These phenomena are frequently discussed in literature pertaining to fear responses, posttraumatic stress disorder (PTSD), sexual assault, Parkinson's disease, and other neuropsychiatric conditions. Research on human tonic immobility often draws upon foundational studies in avian models, such as chickens, which exhibit tonic immobility as a fear-induced response (Marx et al. 2008). The investigation of tonic immobility in humans constitutes a relatively recent and rapidly expanding field, with important implications for psychology and trauma-informed therapeutic approaches (Marx et al. 2008; Volchan et al. 2011, 2017). Human immobility responses, as part of the defensive behavioral cascade, appear to share homologous features with those documented in non-human animals (Volchan et al. 2017). Elucidating the underlying physiological mechanisms of tonic immobility in humans (Volchan et al. 2011) holds promise for the advancement of more effective interventions for individuals experiencing traumatic stress.

6.1 Parkinson's Syndrome: Can Research on Death Feigning in Insects Contribute to Medicine?

Building upon the discussion in Chap. 5, we begin by examining a study in which artificial divergent selection was applied to *Tribolium castaneum* based on the duration of death-feigning behavior. This selection produced two distinct strains—one exhibiting short duration and the other long duration of death feigning—whose genomes were subsequently compared (Uchiyama et al. 2019; Tanaka et al. 2021).

Uchiyama et al. (2019) provided the first evidence worldwide that, in addition to dopamine signaling in the brain, genetic pathways associated with caffeine metabolism, cytochrome oxidoreductase activity, lifespan regulation, and circadian rhythm control—all of which regulate temporal biological processes critical for survival—play a role in modulating death-feigning behavior. This finding represents a significant advancement in elucidating the genomic basis of movement initiation versus immobility. Moreover, these insights offer valuable perspectives on the evolutionary underpinnings of human behavioral responses related to survival and hold potential for future medical applications.

As of autumn 2022, genomic analyses are still ongoing. Readers who have followed the discussion thus far may have observed a notable finding: the long-duration death-feigning strain exhibits significantly reduced dopamine levels in the brain relative to the control strain (Miyatake et al. 2008). Administration of dopamine results in a shortened duration of death feigning and an increase in locomotor activity in these beetles (Nishi et al. 2010). Furthermore, the long-duration strain demonstrates locomotor abnormalities, particularly impairments in turning behavior (Maekawa et al. 2020, 2021).

This phenotype bears resemblance to symptoms observed in a specific human neurological disorder—Parkinson's disease. Indeed, my late mother suffered from

Parkinson's disease during the final years of her life, and having cared for her, I could not help but notice parallels between her condition and the long-duration death-feigning strain of *Tribolium castaneum*.

Motivated by this observation, I requested my collaborators to conduct a comparative analysis of homologs of human Parkinson's disease–associated genes between the short- and long-duration strains (Tanaka et al. under review). Specifically, for *Tribolium castaneum*, the reference genome of a standard, unselected strain (Tcas5.2) is publicly accessible via the NCBI genome database (https://www.ncbi.nlm.nih.gov/). We performed comparative genomic analyses between the short-duration strain and the reference genome, as well as between the long-duration strain and the reference genome. Subsequently, we focused on genomic regions homologous to the amino acid sequences of human Parkinson's disease–related genes to identify potential genetic variations.

Ingenuity Pathway Analysis (IPA, Qiagen), utilizing the human dataset from the knowledge base, was employed to investigate the metabolic pathways associated with these genes.

The analysis detected small genetic variations in 6734 regions in the long-duration (L) strain and 1476 regions in the short-duration (S) strain. Notably, the variations were predominantly enriched in the L-strain, which exhibited a greater number of mutations in Parkinson's disease–related genes compared to the S-strain. In other words, numerous mutations were identified in genes linked to Parkinson's disease within the long-duration strain, which also displays locomotor impairments. This study is currently being prepared for publication (Tanaka et al. under review).

Although caution must be exercised when extrapolating findings from insects to humans, it is conceivable that genetic insights derived from this beetle may ultimately inform strategies for the treatment of human neurological disorders. As I continue to analyze the genomic data, I am often reminded of my mother. We are now entering the era of genome editing.

6.2 Fear and Posttraumatic Stress Disorder (PTSD)

Recent research has demonstrated that humans exhibit freezing behavior as a defensive response even during infancy (Rousseau et al. 2024). In their study, Rousseau et al. (2024) recorded and analyzed video data of 30 infants, including measurements of freezing duration. The authors reported: "The very first human behaviors at birth comprise a cascade of flight and fight active responses, as well as flaccid immobility and freezing passive responses to the stress of birth. Passive behavioral responses may subsequently be followed by trauma-related symptoms." Their findings suggest that prenatal stress can result in inhibited defensive responses after birth.

Posttraumatic Stress Disorder (PTSD), a psychiatric disorder, has a well-established association with tonic immobility (Rocha-Rego et al. 2009; Lima et al. 2010; Volchan et al. 2011; Hagenaars and Hagenaars 2020; Barbano et al. 2025). Tonic immobility, also termed thanatosis, is characterized by temporary motor inhibition and is considered an adaptive response to extreme fear. This phenomenon is extensively documented in animal models and has been observed in humans as well.

In humans, tonic immobility (TI) has been documented primarily in the context of sexual assault. Freezing responses triggered by fear have also been reported in situations involving aviation accidents, naval disasters, and other traumatic events (Abrams et al. 2009). Coimbra et al. (2023) found that trauma-induced TI is correlated with the severity of posttraumatic stress disorder (PTSD) symptoms, occurs more frequently in cases of interpersonal violence, and tends to present with greater severity in women. They further highlighted the necessity for longitudinal studies to better elucidate the role of tonic immobility in the onset and progression of psychopathology.

Moreover, the intensity of TI is thought to correlate positively with the severity of PTSD symptoms (Rocha-Rego et al. 2009; Lima et al. 2010; Hagenaars and Hagenaars 2020). Although Barbano et al. (2025) describe research on the association between PTSD and tonic immobility in humans as a relatively nascent field, they conducted an interview-based study examining TI-related behaviors in women who had experienced sexual assault within the preceding seven years.

6.3 Sexual Assault

Although sexual assault is one form of trauma that can lead to PTSD, tonic immobility (TI) is particularly discussed in the context of rape. Several studies interpret TI as one of the behavioral responses exhibited by victims during forced sexual encounters (Augustyn et al. 2024). Defensive responses to fear in such situations vary depending on the perceived proximity of the threat. When the threat is distant, a freezing response may be elicited; at an intermediate distance, flight attempts may occur; however, when the threat is in immediate proximity—as during an assault—tonic immobility is often activated (Marx et al. 2008).

Tonic immobility, which can manifest as "playing dead" or death feigning, is considered an evolutionarily adaptive survival strategy, as predators tend to avoid prey that appears lifeless (Bovin and Marx 2010). Compared to other trauma types, TI is especially prevalent as a response to sexual assault (Kalaf et al. 2017). Furthermore, Kalaf et al. (2017) reported that peritraumatic tonic immobility is more strongly associated with sexual trauma—particularly childhood sexual trauma—than with other forms of trauma in the general population.

6.4 Mental Disorders: Hysteria, Dissociation, and Functional Neurological Symptom Disorder

One recognized mental disorder is hysteria syndrome, classified as a form of dissociation—an unconscious defense mechanism in which a discrete set of psychological or behavioral processes becomes segregated from the individual's remaining mental functions. In psychological discourse, hysteria refers to a condition characterized by an overwhelming and unrestrained emotional response, which may manifest as a temporary alteration in mental or affective state (Basavarajappa et al. 2020).

Dissociation encompasses a broad spectrum, ranging from normative, transient phenomena observed in the general population to pathological manifestations warranting clinical intervention. Recent hypotheses propose that dissociation may represent a variant of thanatosis, or "death feigning" reflex. Traditionally, stress responses have been conceptualized in terms of Cannon's "fight-or-flight" model, characterized by sympathetic nervous system activation and elevated levels of stress hormones, particularly glucocorticoids. However, emerging theoretical frameworks suggest the existence of a third defensive strategy—immobility or "freezing"—which may operate in parallel with, or as an alternative to, the canonical fight-or-flight response (Komagamine et al. 2021).

Major psychiatric disorders include epilepsy, bipolar disorder, and schizophrenia. Among these, research on epilepsy—historically referred to as hysteria—has led to the hypothesis that reduced function of anti-N-methyl-D-aspartate receptors (NMDARs) may contribute not only to psychosis but also to functional neurological disorders, hypnosis, and other altered states of consciousness. Komagamine et al. (2021) proposed a continuum model linking psychosis with phenomena such as hypnosis, thanatosis (death feigning), near-death experiences, and other functional neurological conditions.

Future research may benefit from identifying homologous genes implicated in human psychiatric disorders and comparing their mutations with those found in death-feigning beetles, potentially providing novel insights into the molecular underpinnings of these complex states.

6.5 Near-Death Experience and Thanatology

An essay also exists exploring the neurobiological mechanisms underlying near-death experiences, as well as the relationship between rapid eye movement (REM) sleep and death-feigning behavior (Nelson 2021). Thanatosis is closely linked to thanatology, the interdisciplinary scientific study of death and dying. Thanatology encompasses biological, medical, forensic, psychological, and social perspectives on the end of life (Anderson 2016; Anderson et al. 2018).

According to Peinkhofer et al. (2021), near-death experiences (NDEs) have been reported across diverse cultures, geographical regions, and historical periods. This cross-cultural and temporal ubiquity suggests that NDEs may possess a biological origin and function. Following a preregistered protocol, we investigate the hypothesis that thanatosis—the death-feigning behavior employed as a last-resort survival strategy in animals—constitutes the evolutionary origin of NDEs. It has been reported that both animals and humans, including humans attacked by "modern" predators, can experience both thanatosis and near-death experiences (NDEs), and that these two phenomena share overlapping phenomenological features and effects. In other words, the available evidence outlines that thanatosis may constitute the evolutionary foundation of NDEs, with a shared biological function of conferring survival benefits. Peinkhofer et al. (2021) further proposed that the acquisition of language enabled humans to transform these events from the relatively stereotyped manifestations of death feigning under predatory threat into the richer perceptual experiences characteristic of NDEs, thereby extending their occurrence to non-predatory contexts.

6.6 Industrial Applications of Tonic Immobility

In the case of the sap beetle, a pest species affecting plum trees, attempts have been made to reduce damage by shaking the trees to dislodge the insects and remove them from their host plants. During these experiments, researchers measured the insects' behavioral responses and the duration of tonic immobility following their fall to the ground. A negative correlation was observed between walking duration and the frequency of tonic immobility and startle responses. It was further suggested that the vibration interval may influence the length of tonic immobility, potentially due to habituation to the stimulus (Kishi and Takanashi 2019). Immobility induced by vibrations in the order Coleoptera has been reviewed by Takanashi and Kojima (2021).

As discussed in Chap. 3, tonic immobility is also a significant phenomenon in marine biology, particularly in the context of tagging sharks and fish for individual tracking. Tag attachment is ideally performed while the animals are in a state of tonic immobility, and the duration of this immobile state has been studied across various species.

References

Abrams MP, Carleton RN, Taylor S, Asmundson GJG (2009) Human tonic immobility: measurement and correlates. Depress Anxiety 26:550–556. https://doi.org/10.1002/da.20462

Anderson JR (2016) Comparative thanatology. Curr Biol 26:R543–R576

Anderson JR, Biro D, Pettitt P (2018) Evolutionary thanatology. Philos Trans R Soc Lond B Biol Sci 373:20170262

Augustyn M, Pinchevsky GM, Li N (2024) What is rape? Elements of rape and application of the criminal label. J Exp Criminol. https://doi.org/10.1007/s11292-024-09633-w

Barbano AC, Pigorr DM, Raugh IM, Ellem JR, Sexton MD, Anderson RE (2025) Posttraumatic stress disorder symptoms, posttraumatic cognitions, and dissociative experiences following rape: the roles of rape tactics and peritraumatic tonic immobility. J Trauma Stress (International Society for Traumatic Stress Studies):1–13

Basavarajappa C, Dahale AB, Desai G (2020) Evolution of bodily distress disorders. Curr Opin Psychiatry 33:447–450

Bovin MJ, Marx BP (2010) The importance of the peritraumatic experience in defining traumatic stress. Psychol Bull 137(1):47–67

Coimbra BM, Hoeboer CM, van Zuiden M, Williamson RE, D'Elia AT, Mello AF, Mello MF, Olff M (2023) The relationship between tonic immobility and the development, severity, and course of posttraumatic stress disorder: systematic and meta-analytic literature review. J Anxiety Disord 97:102730

Hagenaars MA, Hagenaars JAP (2020) Tonic immobility predicts poorer recovery from posttraumatic stress disorder. J Affect Disord 264:365–369

Kalaf J, Coutinho ESF, Viletea LMP, Luz MP, Berger W, Mendlowicz M, Volchan E, Andreoli SB, Quintana MI, Mari JJ, Figueira I (2017) Sexual trauma is more strongly associated with tonic immobility than other types of trauma—a population based study. J Affect Disord 215:71–76

Kishi M, Takanashi T (2019) Tonic immobility and startle responses induced by substrate-borne vibrations in the sap beetle, *Phenolia (Lasiodites) picta* (Coleoptera: Nitidulidae). Jpn J Appl Entomol Zool 63:13–16

Komagamine T, Nokubun N, Hirata K (2021) Hystero-epilepsy in the Tuesday Lessons and NMDA receptor function: a hypothesis for dissociative disorder. Med Hypotheses 150:110567

Lima AA, Fiszman A, Marques-Portella C, Mendlowicz MV, Coutinho ESF, Maia DCB, Berger W, Rocha-Rego V, Volchan E, Mari JJ, Figueira I (2010) The impact of tonic immobility reaction on the prognosis of posttraumatic stress disorder. J Psychiatr Res 44:224–228

Marx BP, Forsyth JP, Gallup GG, Fusé T, Lexington JM (2008) Tonic immobility as an evolved predator defense: implications for sexual assault survivors. Clin Psychol Sci Pract 15:74–79

Maekawa T, Ohara K, Zhang Y, Fukutomi M, Matsumoto S, Matsumura K, Shidara H, Yamazaki S, Fujisawa R, Ide K, Nagaya N, Yamazaki K, Koike S, Miyatake T, Kimura K, Ogawa H, Takahashi S, Yoda K (2020) Deep learning-assisted comparative analysis of animal trajectories with DeepHL. Nat Commun 11:5316. https://www.nature.com/articles/s41467-020-19105-0

Maekawa T, Higashide D, Hara T, Matsumura K, Ide K, Miyatake T, Kimura K, Takahashi S (2021) Cross-species behavior analysis with attention-based domain-adversarial deep neural networks. Nat Commun 12:5519. https://doi.org/10.1038/s41467-021-25636-x

Miyatake T, Tabuchi K, Sasaki K, Okada K, Katayama K, Moriya S (2008) Pleiotropic anti-predator strategies, fleeing and feigning death, correlated with dopamine levels in *Tribolium castaneum*. Anim Behav 75:113–121

Nelson KR (2021) From the stillness of feigning death to near death experience? Brain Commun. https://doi.org/10.1093/braincomms/fcab138

Nishi Y, Sasaki K, Miyatake T (2010) Biogenic amines, caffeine and tonic immobility in *Tribolium castaneum*. J Insect Physiol 56:622–628

Peinkhofer C, Martial C, Cassol H, Laureys S, Kondziella D (2021) The evolutionary origin of near-death experiences: a systematic investigation. Brain. Communications 3:fcab132. https://doi.org/10.1093/braincomms/fcab132

Rocha-Rego V, Fiszman A, Portugal LC, Pereira MG, de Oliveira L, Mendlowicz MV, Marques-Portella C, Berger W, Coutinho ESF, Mari JJ, Figueira I, Volchan E (2009) Is tonic immobility the core sign among conventional peritraumatic signs and symptoms listed for PTSD? J Affect Disord 115:269–273

Rousseau P, Malfilatre G, Matton F, Lecuyer R, Lahaye W (2024) The human newborns' behavioral defense responses to the stress of birth. J Pediatr Neonatol 5(1):1040

Takanashi T, Kojima W (2021) Vibration-induced immobility in coleopteran insects. In: Sakai M (ed) Death-feigning in insects, Entomology monographs. Springer Nature Singapore Pte Ltd, pp 73–92

Tanaka K, Sasaki S, Matsumura K, Yajima S, Miyatake T (2021) Genomic characterization between strains selected for death-feigning duration for avoiding attack of a beetle. Sci Rep 11:1–18. https://doi.org/10.1038/s41598-021-00987-z

Tanaka K, Sasaki K, Yajima S, Miyatake T (under review) Beetle strains artificially selected for longer duration of tonic immobility have more variations in the human Parkinson's disease genomic region

Uchiyama H, Sasaki K, Hinosawa S, Tanaka K, Matsumura K, Yajima S, Miyatake T (2019) Transcriptomic comparison between beetle strains selected for short and long durations of death feigning. Sci Rep 9:14001. https://doi.org/10.1038/s41598-019-50440-5

Volchan E, Souza GG, Franklin CM, Norte C, Rocha-Rego V, Oliveira JM, David IA, Mendlowicz MV, Coutinho ESF, Fiszman A, Berger W, Marques-Portella C, Figueira I (2011) Is there tonic immobility in humans? Biological evidence from victims of traumatic stress. Biol Psychol 88:13–19

Volchan E, Rocha-Rego V, Bastos AF et al (2017) Immobility reactions under threat: a contribution to human defensive cascade and PTSD. Neurosci Biobehav Rev 76:29–38

Yoshikawa T (1989) Sangoku-shi. Kodansha, Tokyo, p 512. (In Japanese)

Future Research on Feigning Death: Prospects and Challenges

7

Abstract

This chapter discusses the current state of research on death-feigning behavior, highlighting both the progress made and the challenges that remain. It is likely that new reports of death feigning will continue to emerge in taxonomic groups where such behavior has not yet been documented. At present, only a limited number of species have been studied in terms of underlying mechanisms, and our understanding of these mechanisms remains superficial. Furthermore, research on the fitness consequences of death-feigning behavior—aside from studies in beetles—remains scarce. One major limitation is the lack of studies conducted under ecologically realistic conditions, where predators coexist with prey. Beyond a few examples—such as studies involving beetles with jumping spiders or assassin bugs, sexual interference in dragonflies, and interactions between water fleas and dragonfly larvae—such research is virtually nonexistent. In addition, no studies have yet explored the phylogenetic relationships and evolutionary origins of death-feigning behavior. This is a critical and necessary area of investigation. Of course, the animal world—including humans—is full of instances of death feigning. This field of research is waiting for you to dive in. Don't move—just come.

Keywords

Perspective · Future work · Phylogenetic tree · Uncharted territory

Research to date has demonstrated that death-feigning behavior is closely associated with the motor activity of the organism. Living organisms generally exhibit two behavioral states: active (locomotive) and immobile. In the active state, an organism may exhibit escape responses such as fleeing or displaying threat postures, whereas

© The Author(s), under exclusive license to Springer Nature Singapore Pte Ltd. 2026

T. Miyatake, *Death Feigning*, Ecological Research Monographs,
https://doi.org/10.1007/978-981-95-5319-8_7

in the immobile state, it engages in tonic immobility or death feigning. Dopamine has been identified as a key neurochemical modulator governing the transition between these behavioral states.

One of the critical research directions for future studies on death-feigning behavior is to elucidate the specific effects of the frequency and duration of death feigning in prey on the behavior of real predators under natural field conditions. It is essential to determine the ecological and behavioral contexts in which tonic immobility is triggered in response to predatory attacks. Key questions include: What intensity or type of stimulus is required to terminate the death-feigning state? Is there an optimal duration of immobility that maximizes survival? What kinds of stimuli are most effective in re-engaging the predator's interest after the prey has adopted a death-feigning posture?

To date, the functional significance of death-feigning behavior has not been thoroughly investigated under fully natural field conditions. This empirical gap is likely attributable to the substantial logistical and observational difficulties associated with conducting such studies in uncontrolled outdoor environments. Paradoxically, the methodological inertia surrounding this issue presents a major obstacle to progress in a research domain that, metaphorically speaking, may itself be exhibiting signs of "tonic immobility."

We have selectively bred for prolonged duration of death-feigning behavior, and an important next step is to determine whether dopaminergic expression differs before and after the expression of tonic immobility. This has often been highlighted as a critical question in the field. However, addressing this experimentally presents a significant technical challenge. When investigating gene expression in insects that exhibit death feigning, it is common to rapidly freeze specimens using liquid nitrogen. However, in individuals who initiate tonic immobility immediately upon tactile stimulation, this handling itself induces the very behavior under investigation. Therefore, the development of a method to freeze-fix such individuals without triggering death feigning is essential for accurate molecular analysis.

From an applied perspective, the potential practical uses of death-feigning behavior should not be overlooked, particularly in contexts that may benefit human livelihoods. For instance, some studies have explored the possibility of exploiting tonic immobility in pest species as a means of pest control. As discussed in Chap. 6, in certain insect pests that damage fruit trees, mechanical vibration of the tree can cause the insects to drop off due to an induced death-feigning response. This suggests a potential association between vibrational stimuli and tonic immobility.

Furthermore, as described in Chap. 3, the analysis of volatile compounds in weevils using olfactometric techniques has been successfully applied to individuals exhibiting death-feigning behavior, indicating that such physiological states do not necessarily preclude chemical analysis. These findings collectively suggest that death-feigning behavior may have future applications in the field of applied entomology, particularly in the development of novel, behavior-based pest management strategies.

In addition, genetic analyses of the genes associated with death-feigning behavior, as well as gait analyses of selectively bred strains exhibiting prolonged tonic immobility, have revealed mechanisms that appear to be conserved across phylogenetically distant taxa. These findings raise the possibility that such mechanisms may have relevance for human neurological conditions, including potential implications for the treatment of Parkinson's disease. Moreover, the genomic-level associations between tonic immobility and psychiatric disorders remain largely unexplored, representing a promising avenue for future research.

As discussed in Chap. 6, defensive behavioral responses such as freezing and tonic immobility are present in humans from infancy and have also been documented in nonhuman primates, including monkeys. These responses may be considered evolutionary antecedents—or preadaptations—to the more specialized behavior of death feigning. This evolutionary trajectory likely involves both predator cognitive mechanisms and deep psychological foundations. Specifically, it may reflect a transition from freezing, which involves minimal movement without a distinct posture, to death feigning, which is characterized by the adoption of a distinct and fixed posture.

Future comparative analyses of immobility-related postures, especially when examined through a phylogenetic lens, have the potential to yield novel insights. Such research could help uncover the evolutionary origins and functional diversity of death-feigning behavior and may ultimately transform our understanding of this complex and understudied phenomenon.

Regarding the prevalence of death-feigning behavior across the biological world, future discoveries in natural history are expected to greatly expand our understanding and present vast opportunities for new research. As of now, the organisms listed in Chap. 3 (Table 3.1) represent only a limited subset of the diversity of life, and even within the class Insecta—comprising 31 recognized orders—death-feigning behavior has been documented in fewer than half. For the majority of insect taxa, such behavior remains entirely unstudied and largely unnoticed by the scientific community.

Thus, the observation of death feigning in any organism outside the currently documented taxa would likely constitute a novel contribution to science and should be formally reported. Moreover, we still have little understanding of the historical and phylogenetic factors that influence the evolution and expression of tonic immobility across lineages. Investigating how evolutionary history has shaped the emergence of this behavior across taxa remains a fundamental and largely untapped area of inquiry.

The author sincerely hopes that this book will contribute, even in a small way, to advancing future research on death-feigning behavior in animals. You are warmly invited to delve into the fascinating world of death feigning. Humanity's scientific exploration of this enigmatic behavior has only just begun. There is still so much to discover—so come along and join the journey.

And now, perhaps you are ready to ask yourself: why do humans feign death? Have you found the answer? (Fig. 7.1)

The animal kingdom is abundant with examples of death-feigning behavior. From the era of small, primitive organisms, the strategy of immobility has been conserved continuously, universally, and through multiple evolutionary lineages. It

Fig. 7.1 Humans engage in thanatosis (playing dead) for various reasons. For example, in my laboratory, students nearing the submission deadline for their theses often "play dead" whenever I enter the laboratory room

is plausible that this ancient tactic has been preserved along the evolutionary trajectory all the way to the human world.

In the end, it may be said that this is what Darwin referred to as instinct (Darwin 1883). I hope that this book will serve as a starting point for understanding death feigning (thanatosis) as an instinctive behavior in animals that may have been preserved through the course of evolution.

Index

© The Editor(s) (if applicable) and The Author(s), under exclusive license to Springer Nature Singapore Pte Ltd. 2026
T. Miyatake, *Death Feigning*, Ecological Research Monographs,
https://doi.org/10.1007/978-981-95-5319-8